Technological Change
in Regulated Industries

Studies in the Regulation of Economic Activity

Technological Change in Regulated Industries

WILLIAM M. CAPRON *Editor*

Papers prepared for a conference of experts,
with an introduction and summary

The Brookings Institution / Washington, D.C.

Foreword

REGULATION BY PUBLIC COMMISSION has typically arisen from
public concern about the effectiveness of a free market in producing and
distributing certain kinds of goods and services efficiently and equitably.
Thus the "short-haul, long-haul" railroad rate differentials in the 1870s,
the death of a senator in an airplane crash in the 1930s, and the thalido-
mide tragedy in the 1960s all strengthened the case for public regulation
and led to further constraints on the behavior of firms in related industries.

Although regulatory commissions are concerned primarily with prices,
profits, and entry into markets, their activities also affect innovation in
regulated industries—sometimes by design, but often as an indirect or
unanticipated consequence of regulatory decisions intended to deal with
some other aspect of industry behavior. Recently, scholars and public
figures have questioned the value of regulation as an instrument of public
policy. They contend that the costs of regulation—particularly those
arising from deleterious effects on technological change—usually outweigh the beneficial effects, if any, on prices and income distribution.
While this extreme view may not be warranted, it nevertheless brings
into sharp focus the need for reexamining the whole idea of regulation by
public commission. Research scholars must play an important role in this
assessment, for the existing theoretical models and empirical information
are not adequate for a complete and systematic appraisal.

This volume is designed to contribute to the public debate by advancing
knowledge of one of the most important aspects of regulation: its effects
on technological progress in regulated industries. How seriously do regulatory agencies regard technological progress as a criterion for assessing
the performance of regulated firms? Have regulatory agencies developed
effective procedures for encouraging innovation in regulated firms? How

should the objectives and practices of regulatory commissions be altered to make regulation better serve the public interest?

Some of the most important regulated sectors are examined in detail: electric power, telecommunications, civil air transport, and surface transport. In Chapter 1, William M. Capron introduces the subject and discusses several key issues. Chapters 2 through 6 are adapted from papers that served as the basis for a conference on the regulated industries held at Brookings in February 1969. Chapter 7 summarizes the papers and the main points that emerged from the conference discussion.

The conference participants, listed on pages 227–28, included academic specialists in the field of regulation, businessmen from regulated industries, and officials of federal regulatory agencies. They contributed to this volume in a number of ways, and the editor is particularly grateful to those who made detailed comments on drafts of his introduction and the summary chapter. He is also grateful to Merton J. Peck for his generous counsel and to Joseph A. Pechman, director of the Brookings Economic Studies Program, for his leadership and encouragement. Mr. Shepherd expresses thanks for the help he received from Donald I. Baker, Manley R. Irwin, Leland L. Johnson, Lionel Kestenbaum, Harvey J. Levin, Frederic M. Scherer, Peter O. Steiner, Robert S. Thorpe, and Harry M. Trebing. Mr. Phillips is similarly grateful to Morris A. Adelman, George Eads, Aaron J. Gellman, Robert Perry, and Oliver E. Williamson. Some of the work underlying Mr. Westfield's paper was supported by a National Science Foundation grant to Vanderbilt University.

Mr. Capron, a senior fellow in the Brookings Economic Studies Program at the time of the conference, joined the faculty of Harvard University in the fall of 1969. As the editor, he is greatly indebted to Leonard S. Silk and to Roger G. Noll of Brookings, who performed many tasks that would otherwise have been his. He is especially grateful to Mr. Noll, who assisted in the final editing and collaborated in writing the summary chapter. The editor and authors also thank Evelyn P. Fisher, who checked the manuscript for accuracy of sources and data; Elizabeth Cross, who prepared the manuscript for publication; and Helen Eisenhart, who prepared the index.

This volume and the conference that preceded it were undertaken as part of the Brookings program of Studies in the Regulation of Economic Activity. This study was supported by grants from the Ford Foundation and the Alfred P. Sloan Foundation. The opinions expressed by the editor and authors do not necessarily reflect the views of the trustees, officers, or staff members of the Brookings Institution, the Ford Foundation, or the Sloan Foundation.

KERMIT GORDON
President

July 1970
Washington, D.C.

Contents

TABLES

FIGURES

CHAPTER ONE

Introduction

William M. Capron

ONE OF THE MOST IMPORTANT questions raised about any economic activity is how well it provides for, and adapts to, technological change. From the standpoint of public policy, this question has special importance for industries that operate under close government regulation. What effect does regulation have on the pace and pattern of technological change in such key sectors as transportation, energy production, and communications? Can a common pattern of interaction between technological change and regulation be identified? Are there significant differences among regulated industries, and, if so, do they result from variations in the regulatory approach, from the organization of the private firms, from the nature of technology, or from some other factor?

These questions constituted the agenda for a conference of experts—mostly academic economists—held at the Brookings Institution in February 1969. Preliminary versions of the papers in this volume served as the starting point for the discussion. The summaries of the individual papers (in Chapter 7) focus on these issues, drawing on the discussion at the conference to illustrate the views that emerged. This chapter presents a synthesis of the subject, relying heavily on the papers and the conference but reflecting only the viewpoint of the editor.

As it is used here, "technological change" includes individual, identifiable changes (referred to as "innovations"), as well as the changes that occur year-in and year-out and emerge as rising productivity, measured by such indices as output per man-hour or per man-year. Thus, technological change is not confined to productivity changes in physical equipment. Innovation is a narrower concept included in technological change. Innovations are those distinct changes in technique associated with new

1

plant and equipment (for example, larger rail cars or nuclear power plants) that are introduced by the explicit decisions of management.

Characteristics of Regulation

The regulated sectors are those that come under the jurisdiction of a federal independent regulatory agency created by Congress. The first such agency—the Interstate Commerce Commission (ICC)—was established in 1887 to regulate the railroads. As other sectors have been identified as requiring control beyond the general laws applying to all business activity, Congress has followed the ICC precedent. Regulatory commissions have been established to oversee the electric utility industry and the interstate natural gas industry (Federal Power Commission—FPC), the communications industry (Federal Communications Commission—FCC), and air transport (Civil Aeronautics Board—CAB). The regulation of interstate trucking and of inland water transportation was lodged in the ICC in the 1930s and 1940s, respectively.

While the powers and procedures of the regulatory agencies vary, their regulatory approaches have attributes in common. First, a firm must be licensed by the regulatory agency controlling the activity in which it wishes to engage. Thus entry into the industry is controlled. Industry structure is also controlled to the extent that mergers of regulated firms must be approved by the regulatory agency. The agencies have considerable power over the quality, frequency, type, and location of service provided. For example, once a service has been initiated, permission is often necessary before it may be withdrawn.

Second, the prices charged by regulated firms are subject to approval by the appropriate regulatory commission. While the commissions have differed greatly in the detail and frequency of their price reviews, firms in all the industries considered here are subject to some degree of price regulation. Not only are rate levels controlled, but in some industries rate structures are subject to detailed review.

Third, although the primary focus here is on federal regulation, the activities considered are subject to state and local regulation as well. Generally speaking, state-level regulation operates on the same principles as does federal regulation and presumably has much the same impact.

Regulation is widely held to have had an adverse effect on technological change in at least some of the industries considered, though this view lacks an accepted analytical and empirical base. But an apparent paradox is also recognized—if regulation has inhibited the pace of innovation, why have all the regulated industries enjoyed long-term productivity increases that are above the national average (and certainly higher than those in most manufacturing industries)? Besides above-average general gains in productivity, there have also been many dramatic innovations. If regulation is supposed to lay the "dead hand of inertia" on the industries under its jurisdiction, how can this be explained?

There are a priori reasons for believing that regulation of the sectors considered at the conference—communications, electric utilities, surface transportation, and air transport—has had an effect on the pace and pattern of technological change in these industries. Regulation imposes constraints on the freedom of action of the managements of regulated firms; among the important management decisions so constrained (directly or indirectly) are those having to do with the introduction of new technology. Unfortunately, as the chapters in this volume and the discussion at the conference show, the a priori expectations cannot easily be tested against the record because (1) regulation is only one of many factors that account for the actual course of technological change in these industries; (2) the industries themselves differ in significant respects; and (3) the actual record of technological change in any industry does not, in and of itself, indicate what changes could or would have taken place without regulation.

Characteristics of the Regulated Industries

Many of the regulated industries show significant economies of scale; that is, unit costs fall as the scale of production increases. They are also relatively capital intensive; that is, capital is a large proportion of total inputs. Indeed, the rationale for regulating a particular industry is often based on the fact that it has these two characteristics. In telephonic communications, and to a lesser extent in rail transportation, both characteristics are very pronounced. These industries are sometimes called natural monopolies, since their scale economies make competition inefficient for society as a whole. And in some of these industries, unregulated competition would lead to monopoly by a single firm simply because the largest firm can have the lowest costs and drive its rivals from the market.

In air transport, and especially in the trucking industry, economies of scale are not very significant. The detailed discussion of the regulation of these sectors reveals that the relative unimportance of both scale economies and capital intensity has influenced the way in which the regulatory process has affected the pace of technological change.

Two other characteristics that are significant in several, though not all, of the regulated sectors are rapid growth and extensive research and development (R&D). The steady and well-above-average growth of demand (in all except the railroad industry) has apparently been caused by, and also had an effect on, technological progress. As to cause, the rapid growth of demand has been in part a result of the continuing fall in the relative prices of the services provided. Improved technology (together with scale economies) has played a major role in making possible these relative price reductions. As for the effect, the steady growth in demand has encouraged a rapid investment in plant and equipment which usually incorporate the new technologies.

The regulated industries tend to be "high technology" industries. They have benefited from comparatively large investment in R&D—though, as will be noted below, most of the R&D affecting them is not undertaken by the regulated firms, except in the communications industry.

Quite apart from any effect of regulation, it would appear that industries which have all four of these characteristics—most clearly exemplified by telephone communications—could be expected to have a relatively high rate of increase in productivity and to have introduced many innovations.

Effects of Regulation on Innovation

While regulation is often regarded as a negative factor in technological change, the regulatory forces do not all operate in the same direction. Regulation has many effects, some leading to more and some to less technological change; and each factor works out differently in different industries. The different effects that various aspects of regulation appear to have on the pace and pattern of technological change are considered below.

The Averch-Johnson Effect

Considerable attention has been given to the view that the usual mode of regulation in the United States systematically biases management decisions of regulated firms in the direction of investing more heavily in plant and equipment than they would if they were unregulated. Regulation may also bias firms toward developing and using more capital-intensive production processes than they otherwise would. This thesis was developed most clearly in a paper published in 1962 by Averch and Johnson.[1] Although others have advanced this idea of overinvestment, or capital-intensive bias, it is usually identified as "the Averch-Johnson effect."

Regulatory commissions ordinarily determine a "fair rate of return" which firms under their jurisdiction will be permitted to earn. Total profits will then be a function of this rate of return and of the amount of capital invested in the regulated activity—the "rate base" recognized by the regulatory agency. The larger the rate base, the greater profits will be, and the higher the investment in plant and equipment, the greater the base. Therefore, according to Averch and Johnson, management has an incentive to use capital-intensive methods to provide its service. Some argue that one result of the Averch-Johnson effect is to encourage innovation, since the introduction of new technology frequently requires adding to the stock of plant and equipment.

Regulated firms may also have an incentive to increase investment by reaching out to new markets, even if these markets are not directly profitable. This is encouraged by the blessing all regulatory agencies give to the practice of price discrimination. Thus the regulated firm, which may have a strong monopoly position and face a relatively inelastic demand in certain markets, can raise prices in those markets, thereby earning enough to cover low returns or losses in markets where it may face more competition from alternative services and where demand is therefore relatively more elastic.

The extent to which firms have in fact acted in response to the investment incentives provided by rate-base regulation has yet to be clearly demonstrated.

The Regulatory Lag

A commission's response to changes in the situation of the regulated firm—for example, a change in its costs—is typically very slow. The commission sets the level and pattern of rates that a firm under its jurisdiction may charge so that the net income the firm can be expected to

1. Harvey Averch and Leland L. Johnson, "Behavior of the Firm under Regulatory Constraint," *American Economic Review,* Vol. 52 (December 1962), pp. 1052–69.

earn will represent a fair rate of return on its investment. If the firm subsequently increases its net income by introducing a cost-reducing innovation, it can enjoy higher earnings until the regulatory agency gets around to reexamining its rate level and rate structure. At the same time, when increasing costs reduce the net income of a regulated firm, it may be forced to ask for a rate increase. Firms are ordinarily anxious to avoid going before the regulatory commission for any reason, since the proceedings are often drawn out and costly to management in time and energy. (Indeed, the regulatory lag is accounted for in large part by the time that elapses between initiation of a proceeding and final action.) A request for a rate increase raises the possibility, unappealing to management, that the regulatory agency will require an adjustment in the structure of rates. Thus, it is argued, the regulatory lag provides an incentive to introduce cost-reducing innovations.

If one accepts the view that firms are reluctant to go before the regulatory commission, especially to ask for rate increases, it is plausible to expect that the incentive to make cost-cutting innovations will be particularly strong during periods of general inflation. A regulated firm may be able to offset inflation-induced increases in wage rates and in the prices of other purchased inputs by introducing cost-reducing changes into the production process. To the extent that a firm is successful in this endeavor, it can postpone or eliminate the need to ask for an upward adjustment in rates. On the other hand, regulated firms have other reasons for introducing cost-reducing innovations. Thus, once again, it is difficult to get hard evidence to support this line of argument, and none has yet been developed that suggests its relative importance.

System Integrity

The need for system integrity is sometimes acknowledged explicitly by regulatory agencies and, when it is to their advantage, by the regulated firms; in other cases, it is implicit in the constraints placed on management action. The forces working for system integrity come not only from regulators but also from the special character of the processes by which certain regulated services are produced and delivered.

Telephonic communication is the most obvious example of an industry in which the public interest calls for some degree of system integrity. Equipment used in the system must permit signals to be transmitted between any two users without degrading or disrupting signal transmission for other users. This necessitates the ability to connect two telephones temporarily, which in turn means that each local exchange must be tied, directly or indirectly, to all other local exchanges. In other words, users will enjoy the full service that current technology permits only if the total system is integrated. While there is no *technical* reason why a single firm must operate the entire system, there seems little doubt that the regulatory procedures for ensuring system integrity are much easier because so much of the telephonic communications system is operated by a single firm, American Telephone and Telegraph (AT&T). (This applies especially to the one area where AT&T does have a virtual monopoly—the so-called long-lines division, through which individual local exchanges are linked.)

Another industry in which system integrity plays a role is railroad transport. Since efficiency requires that rolling stock move from one railroad line to another, such equipment as track gauges, coupling devices, and so forth, must be compatible. But unlike the telephone industry, which is dominated by a single firm, the railroad industry is made up of a number of separate firms, and the railroad regulatory agency must actively police the system to ensure integrity.

The different industry structures, in large part controlled by the regulatory agencies, create different environments for technological change. In communications, AT&T can be permitted great latitude in introducing new technology, since the problem of system integrity is largely internal to the firm. (As noted below, AT&T has in fact invoked the need to maintain system integrity to justify its control over equipment tied into its systems. Thus system integrity may be used as an argument for maintaining and extending monopoly and for discouraging or preventing innovation. It has also been used as an argument for certain mergers in the transport field.) On the other hand, innovation may have been inhibited in railroad transportation because of the need for coordination among independent firms when certain kinds of new equipment are to be introduced. Certainly the regulatory agency has a greater responsibility in this regard in railroading than in communications. Since the ICC does not have direct control over the amount and type of equipment used, its policies on mergers and other actions of the railroads are especially important from the standpoint of system integrity. To the extent that the commission encourages system integration in railroading, it may also promote innovation.

Risk Reduction

While the regulatory agency cannot guarantee a firm against loss, it ordinarily acts to protect firms under its control from the effect of "mistakes." For example, if a regulated firm tries a new kind of equipment, it is permitted to include the cost of this equipment in its rate base (assuming it is kept in service), even if the equipment turns out to be less efficient than that which it was designed to replace. If the firm continues to use the new, less efficient equipment, the incentive to invest will have a perverse effect on efficiency. The identification by members of regulatory commissions with "their" industry, plus their commitment to see that the public is well served by it, often leads them to underwrite whatever risky ventures their firms choose to undertake.

Prevention of Monopoly Profits

Joseph A. Schumpeter argued persuasively that the incentive to earn very large profits spurs entrepreneurs to introduce new techniques.[2] Despite the regulatory lag discussed above, the commissions do their best to prevent firms under their jurisdiction from earning high monopoly profits. Managements of these firms, therefore, do not have the incentive to take high risks that the genuine entrepreneurs of the Schumpeterian model would take.

It is alleged that this regulation-induced conservative bias has led regulated firms to avoid large, high-risk ventures. For example, the electric utility industry supports little research and development. It has been wholly dependent on equipment suppliers and, more significantly, on the Atomic Energy Commission for nuclear power, the major innovation in this industry in the past several decades. Another example is the commercial air transport industry, which has shown no eagerness to support aviation research and development. The supersonic transport program, if it goes forward, will need substantial federal support. The air carriers have apparently been unwilling to underwrite this high-risk, high-cost venture.

2. *The Theory of Economic Development: An Inquiry into Profits, Capital, Credit, Interest, and the Business Cycle* (Harvard University Press, 1934), especially pp. 132–53.

In short, regulation tends to cut off both the upper and lower ends of the profit-possibility distribution a firm faces. On the one hand, the regulated firm is protected against the risks of loss inherent in technological change; on the other, it is denied the supernormal profit that the unregulated, successful innovator can expect to earn. The net effect of these two offsetting influences on management's propensity to innovate is unclear.

Restraints on Entry

Regulated industries are closed systems. Once they are established, regulated firms face practically no direct threat to their positions in the market. Rarely is a franchise, license, or certificate shifted from one firm to another without the consent of the original holder. Regulatory agencies, moreover, are usually very cautious in authorizing new entry. Neither the ICC nor the CAB will grant a certificate for an additional trucker or airline to operate on a given route until convinced that the volume of traffic is sufficient for the existing carriers and the new entrant as well. In issuing additional certificates for an existing route, regulatory agencies generally give strong preference to firms already operating in the industry. Thus regulation tends to discourage the entry of new, innovative firms.

Even when not protected by real economic barriers to entry (as, for example, when there are large economies of scale), regulated firms nonetheless operate in a protected, closed-system atmosphere. They are not usually threatened by aggressive new firms entering an industry on the basis of a service-improving or cost-reducing innovation. Where limited competition is permitted by the regulatory agency (as in commercial aviation and motor carriers), an innovation introduced by one firm may initiate strong pressure on those operating in the same market to imitate quickly. Thus the regulatory framework, which ordinarily suppresses competition and does not positively promote technological change, encourages the rapid *diffusion* of technological change once it occurs. This point will be considered further in connection with commercial aviation.

The conservative bias that regulatory agencies have against permitting —let alone encouraging—new entry arises in part from the commonly held view that the public interest is best protected by maintaining stable, low-risk industry structures, in which the quality and frequency of service, once established, will be maintained. Regulators also act to shield from losses firms already established in a regulated activity. This too leads to bias in favor of protecting the status quo.

The best example of the conservative bias is provided by the ICC, which is responsible for regulating competing modes of transport. The commission has not developed policies and procedures to compensate a railroad that loses profitable traffic to a motor carrier offering a new and better service. As a consequence, the commission has been slow to approve new service made possible by new and improved equipment. For example, the whole history of the ICC's actions in overseeing the introduction of piggybacking in freight transport supports the view that an innovation-retarding bias has had a significant effect on technological development in the industry. The commission is understandably loath to approve changes that threaten existing operators, especially since the railroads generally have been earning below-average profits.

This tendency is reinforced by the regulatory process itself, because the firms that would be hurt by a change have, in the regulatory agency, a forum in which to plead their cause. Not only other carriers (in the same or competing modes) but also localities and shippers may complain to the ICC that a proposed change would mean damage or loss to them. For example, railroads have been impeded in their attempts to cut costs by discontinuing unprofitable service and abandoning particular communities altogether. Unquestionably such changes may affect existing firms and industries that have comparative locational advantages. It is not surprising that regulators hesitate to approve changes that will upset existing relationships among communities and business firms. Without any means of facilitating adjustments and, where necessary, compensating those adversely affected, regulators give considerable weight to the distributional effects of proposed changes, and tend to resist and even reject innovations that have such effects to a significant degree.

Summary

The net impact of various forces operating through the regulatory process on technological change and on specific innovations is complex and challenging. No assessment that would apply to all of the regulated sectors seems possible on the basis of evidence so far available, partly because of differences in the basic characteristics of these industries and partly because of differences in approach of the various regulatory bodies. The difficulty of making a generally applicable assessment is illustrated by the contrast between surface transportation and communications.

Regulation in railroad and truck transport has almost certainly slowed and distorted the pace and pattern of technological change. As noted above, the ICC has been preoccupied with protecting the interests of the carriers and users. Also it has been constrained by the need to protect and promote system integrity. And finally, it has virtually complete authority to review and order changes in carrier rates.

In contrast, the net impact of regulation on the pace and pattern of technological change in telephonic communication has probably been positive, or at worst neutral. The structure of the industry, completely dominated by a single firm that is both horizontally and (more significantly in this context) vertically integrated, is the most important factor in explaining this. The main influence arising from FCC regulation (if there has been any) on technological innovations has apparently been the Averch-Johnson effect discussed above. Until very recently the FCC has supported AT&T in the latter's insistence that system integrity requires that AT&T retain extensive control over equipment linked to the telephone network. This has protected and encouraged AT&T innovations, although it has discouraged and inhibited innovation by other firms. On balance, it is not clear how this particular posture of the FCC has affected the rate of innovation—and of technological change more generally—but it seems likely that it has biased the direction of R&D at the Bell Laboratories and the kind of innovation introduced by AT&T. Naturally AT&T wants to preserve and extend its dominant position in the telephone communications market and will therefore try to introduce new technology that requires it to provide and own as much of the equipment as possible. For example, it will prefer equipment that, to be used effectively, must be closely integrated with other parts of the system.

To assess the net impact of regulation on technological change is more difficult in the electric utilities and civil aviation industries than in surface transport or in telephone communications. One of the clear-cut differences among the industries considered here is the role of federal government support of R&D as a source of new technology. Consider, by way of contrast, railroad transport and civil aviation. Virtually no federal support has been given to R&D in the former. The only recent exception is support for improved passenger transportation under the Northeast

Corridor Program administered by the Department of Transportation. Such innovations as have been introduced have come almost exclusively from the producers of railroad equipment.

In aviation, the federal government has been the major single source of funds for R&D. Moreover, the government itself has undertaken substantial R&D at its own facilities, both civil and military, from which a number of significant innovations have developed. Commercial air carriers have done almost no research, nor have they ordinarily contracted with others to undertake it. Major technical advances have usually come from efforts supported by the military to improve military aircraft capabilities. The specific application to civil aviation of new technology developed through military R&D has been supported by major aircraft producers. In some cases, civil aircraft have been relatively close copies of military ones; in others, new civilian aircraft have been developed that do not have a close military counterpart but that incorporate various innovations developed in government-supported programs.

The following five chapters present the papers that served as the basis of the conference discussion. The versions given here are revisions of the original papers circulated to conference participants, based in part on the conference discussion. In Chapter 2, Fred Westfield presents a theoretical analysis of the impact on innovation of several regulatory techniques. He demonstrates the strengths and limitations of theory as applied to important policy issues. In Chapters 3 through 6 technological change in important sectors of the economy is explored, and observed interaction between change and regulation in those sectors in recent decades is considered. William Hughes examines the electric power-generation industry; William G. Shepherd focuses on common carrier communications; Almarin Phillips traces the introduction and use of new aircraft by commercial carriers since 1932; and Aaron Gellman examines a number of specific ways in which regulation has had an impact on technological change in various surface transport modes. The final chapter presents a summary of the discussion at the conference and suggests a few general conclusions.

CHAPTER TWO

Innovation and Monopoly Regulation

Fred M. Westfield

THE STUDY OF THE RELATION between regulatory policy and innovation is fraught with difficulties. Innovative processes are not wholly understood by economists. Theory about the determinants of the rate and direction of technological change is at a primitive stage. And if one chooses a broader definition of innovation, theory is even less satisfactory. But this is only half the problem, for the realm of the positive theory of regulation is also a wasteland. Thus a theoretical paper attempting to join these two topics must be highly tentative and exploratory.

The approach followed here is to analyze the traditional economic theory of the firm as it applies to regulation and to technological change. One may imagine a firm subject to regulation selling a product or a service, under monopoly conditions, in one or more markets. The firm faces one or more downward-sloping demand curves—more than one if it sells in spatially or temporally separated markets or is able to "differentiate the product" and engage in price discrimination. Current technological possibilities, the structure of markets in which the firm must buy, and relative input prices determine a minimum level of cost for each configuration of production and sales. Thus, corresponding to every set of quantities of the firm's output is an associated level of maximum total (unregulated) profit. The relation between the maximum attainable profit and output is represented by a profit-possibility schedule or function. Economic theory usually postulates that a profit-possibility function has a well-defined maximum. The curve in Figure 2-1 shows this relation for a firm producing a single product or service (or a fixed-proportion bundle of several products or services). It is applicable both to a firm selling in

13

Figure 2-1. Profit-Possibility Curve for Firms Producing Single Products

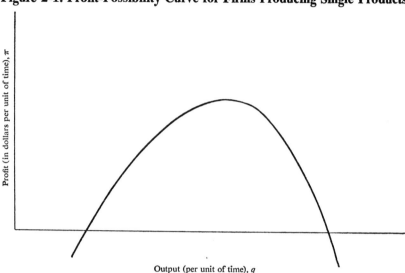

various markets at different prices and to one producing at various locations with different production and transport costs. Points on the curve assume that any market segmentation of sales and any regional dispersion of production are undertaken when it is profitable to do so. When there are many products or services, a two-dimensional figure cannot show the profit opportunities; instead a general mathematical relation is necessary.

It is assumed in this chapter that, in the absence of regulation, firms will move to the peak of the profit-possibility function. Some economists in recent years have wanted to reformulate this classical theory of the firm by stressing managerial goals other than profit maximization. However useful such reformulations may be for some purposes, this chapter is based on the thesis that the profit-maximizing hypothesis explains much of the observed behavior of monopolistic business firms. Although the manager of a monopoly may want "the quiet life," he may be denied this luxury by those who could and would offer stockholders a higher valuation of their equity.

In principle, regulation by government can take many forms. One can imagine rules and regulations so complete that regulation would amount to management, or so weak that regulation would have no effect at all. Generally, a consequence, if not an intent, of regulation is to superimpose on the set of *possible* profit-output combinations shown in Figure 2-

1 a set of *permissible* profit-output points. If the firm subject to regulation has a monopoly position, the economist would want regulation to induce the firm to expand its production and sales, because from the point of view of economics the basic adverse effect of monopoly is in the scarcity it produces. The economist would want the peak of a superimposed profit-permissibility curve to occur to the right of the peak of the profit-possibility curve in Figure 2-1. While the economist's focus is, or perhaps should be, primarily on the horizontal coordinate at which the peak of the profit-permissibility curve occurs, it sometimes seems as if public officials who are responsible for regulation focus more on the vertical component—on the height of the peak. They ask, Are profits "fair"? Are they "reasonable"? In fact, an often-heard charge is that regulators serve the interests primarily of the regulated firms and that they, in truth, organize and enforce what amount to cartels. Such regulation would present a profit-permissibility curve superimposed on Figure 2-1 that would have a peak higher than that of the profit-possibility curve and at a lower level of output—this would be the picture for the fortunate firm with the franchise. Other firms would have profit-permissibility curves with peaks at negative values for the vertical coordinate, or they would simply be prohibited from operating. But such perverse regulation, even though some might argue that it is in fact the norm in the American economy, is not the subject of this analysis.

Regulators could use many policy measures to modify a profit-possibility schedule. The effects of these differ not only in where and how high the peak is, but also in how the underlying cost and revenue structures are altered. Every rule and regulation affects the profit-permissibility schedule, often differently than is intended by the regulators. The analysis to follow will focus on three idealized regulatory techniques: (1) limitations on the rate of return on capital, (2) restrictions on markup, and (3) specification of ceiling prices. The plan is to study how innovative behavior will alter the profit-permissibility curve associated with each of these regulatory techniques and then to use these models to obtain insight into incentives for such behavior.

Profit-Permissibility Curves

How, exactly, is the equilibrium configuration of the monopoly firm influenced by regulatory instruments? Since regulatory constraints are often inequalities, there are ranges of values for the regulatory parameters such that regulation, for the time being, has no effect on the behavior of the firm. The profit-permissibility function will then coincide with the profit-possibility function without regulation close to the latter's maximum, although elsewhere the two may diverge. For the simple monopoly firm, this occurs when a regulatory commission sets an allowable rate of return, an allowable markup on cost, or an allowable price higher than the rate of return, markup, or price that would be associated with ordinary unconstrained profit maximization. Although some economists[1] have tried to argue that this is the typical regulatory situation, such cases —on the borderline of what was previously referred to as perverse regulations—are also of no interest for the purposes of this chapter. The monopolist would be, in truth, unregulated.

Situations where the parameters of the regulatory constraints are of magnitudes that increase the output and lower the prices of the monopolist are examined here. Since the unconstrained monopolist always operates in an elastic region of his average revenue curve, it is appropriate to begin with the assumption that average revenue is elastic.[2] Then it is indeed possible for any of the three regulatory instruments to induce some increase in output and reduction in price. To provide standards for comparison, suppose that each of the regulatory constraints induces the regulated firm to produce the same target output and the same target average revenue (price) by an appropriate selection of parameters. The profit-permissibility curve associated with each of the constraints thus has its peak at the same quantity coordinate. In other words, regulated output and price are assumed to be invariant. Only the rules for getting there and their side effects differ.

In Figure 2-2, the target output of a regulatory commission is denoted by the quantity, q^*. Associated with it is the target price, p^*. The lower quantity of output, q^0, which is associated with the higher price, p^0, would be the equilibrium output for the firm if it were completely unregulated. The curve labeled CP shows the profit-permissibility curve for ceiling-price regulation.

To understand how the ceiling-price profit-permissibility relation is de-

1. See, for example, George J. Stigler and Claire Friedland, "What Can Regulators Regulate? The Case of Electricity," *Journal of Law and Economics,* Vol. 5 (October 1962), pp. 1–16.

2. With price discrimination or regulatory requirements that specify the price differentials between markets, an elastic average revenue can be consistent with inelastic demand.

Figure 2-2. Profit-Permissibility Curves for Three Types of Regulation (Markup, Ceiling Price, and Rate of Return) in Relation to Profit-Possibility Curve

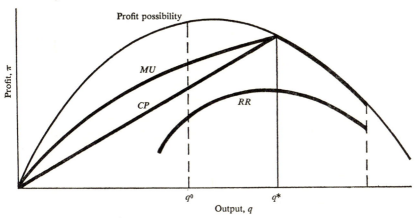

CP =ceiling-price regulation
MU =markup regulation
RR =rate-of-return regulation
q^0 =equilibrium output with no regulation
q^* =target output

rived, one must examine Figure 2-3(a), a conventional cost-demand diagram for the monopoly firm. Here the ceiling-price constraint is designated by the broken horizontal line. The firm may do anything it wants as long as price is equal to or less than p^*. What does this reveal about the permissible profit locus? For levels of output lower than q^*, the price must still be p^* or lower; for example, at output Oa the maximum permissible profit is the rectangle *nmst*. Shortages and deficiencies in the quality of service are created; the amount supplied, Oa, is less than the amount demanded, q^*. Obviously a price lower than p^* or average costs because of inefficiency or padding higher than *am*, while permitted by the regulatory authority, would simply decrease the realized profit associated with output Oa. In the absence of a ceiling price, the output Oa could yield the larger profit *nmvu*.

How do permitted profits vary as output Oa is increased? Evidently, for values of output less than q^*, the area *nmst* increases. If, as in Figure 2-3(a), average costs neither rise nor fall with output, then permissible profits rise in proportion to output, and the relation between permissible profits and output is a straight line through the origin. The profit-permissibility function will rise at an increasing rate if average costs decrease with output; it will rise at a decreasing rate if average costs increase with output. For output less than the target level q^*, the profit-permissibility

Figure 2-3. Cost and Demand Curves for Monopoly Firms

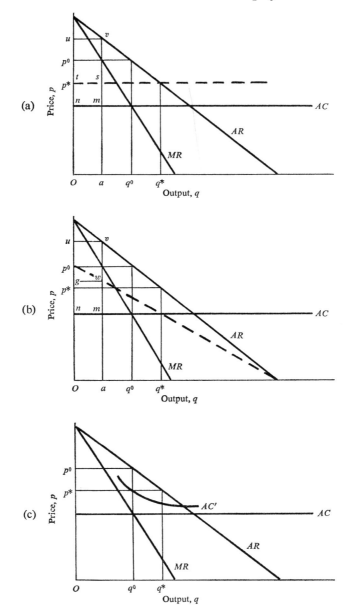

AC = average cost of unregulated firm
AC' = average cost with rate-of-return regulation
AR = average revenue
MR = marginal revenue

p^* = target price
p^0 = price with equilibrium output
q^* = target output
q^0 = equilibrium output with no regulation

function is determined entirely by the structure of costs. For output exceeding the target, the profit-permissibility function coincides with the profit-possibility function since increased output requires prices below the ceiling.

The profit-permissibility curve for markup control is shown in Figure 2-2 by the curve MU. Figure 2-3(b) shows how the curve is constructed. For any price that might be charged, unit costs must equal or exceed a certain level, which depends on the size of the permitted markup. The broken line in Figure 2-3(b) illustrates such a locus of minimum unit costs that must be incurred at various price levels. Its ordinates bear a fixed ratio to the ordinates of the average revenue curve. Thus, if output Oa is sold at a price Ou, unit costs must be at least $Og;$ and total profit is $gwvu$. Since the lowest cost of producing output Oa is $Oamn,$ there should be no difficulty in incurring higher cost through inefficiency: combining inputs in the wrong proportions, buying more of them than necessary, or paying more for them than the going price. Is $gwvu$ the largest profit permitted on output Oa? One other possibility would be to charge a lower price for this output, thereby creating shortages and shedding load as under price control. While costs could be proportionally lower at output $Oa,$ revenues would fall by the same proportion, and realized profits would thus have to be less. It follows that for any output less than q^*, the MU curve is higher than the CP curve; $gwvu$ is necessarily larger than $nmst$. For the particular markup selected, the profit-permissibility curve in Figure 2-2 reaches a peak at q^*, and for any outputs greater than that, the MU curve will coincide with the CP and profit-possibility curves. Note that the shape of the MU curve for output less than q^* is determined quite independently of cost conditions.

The profit-permissibility curve for rate-of-return regulation, RR in Figure 2-2, has somewhat different properties from those of the CP and MU curves. Given the specification of the rate base and the appropriately selected allowable rate of return, the peak of the RR curve also occurs at q^*. However, the total profit that can be earned with output in the neighborhood of q^* must be lower than the profit-possibility curve—and lower than the other two profit-permissibility curves—would indicate for this output. This is because of the Averch-Johnson effect.[3] Effective rate-

3. Harvey Averch and Leland L. Johnson, "Behavior of the Firm under Regulatory Constraint," *American Economic Review,* Vol. 52 (December 1962), pp. 1052–69. See also Fred M. Westfield, "Regulation and Conspiracy," *American Economic Review,* Vol. 55 (June 1965), pp. 424–43.

of-return regulation leads to a substitution of inputs that are included in the rate base for those that are not. Output q^* is not produced at the lowest possible cost. In Figure 2-3(c), one can see only some traces of the phenomenon.[4] The AC' curve shows the least cost that can be incurred at each level of output if it is sold at the maximum average revenue obtainable (as shown by the average revenue curve), and if the factors are combined in such a way that the allowed rate of return will not be exceeded.

It should be noted that the position of AC' is dictated by the position of the average revenue curve. It is higher than AC because rate-of-return regulation, if it is effective, forces an otherwise uneconomical substitution of inputs that appear in the rate base for the cost items that may be subtracted from gross revenue but do not appear in the rate base. How high will AC' be relative to AC? This depends on the allowed rate of return. The lower the allowed rate, the higher is AC'. If the allowed rate were even less than the going rate of interest, AC' would be higher than AR everywhere, and the regulated firm would want to leave the industry. If the allowed rate of return were higher than the rate of interest, then the lower the allowed rate, the higher would be the magnitude q^* that maximizes profit. And for sufficiently flexible technology, a lowering of the allowed rate could shift the peak of the profit-permissibility curve to any output between q^0 and the level where average revenue becomes inelastic, that is, where MR becomes zero.[5]

The particular curve AC' shown here is the one associated with that magnitude of allowed rate of return that yields a constrained profit maximum at q^*.[6] No target output level q^* whatever can be attained simply by making the proper choices for the values of the regulatory parameters. In fact, there are important restrictions on the value of q^*. The most obvious is that the peak of the profit-permissibility curve not be negative, since if it were, the firm would try to leave the industry. But actually the limitations are even more restrictive. Neither with rate-of-re-

4. For a fuller graphical treatment, see Westfield, "Regulation and Conspiracy."

5. For more detail and analysis of the limitations, see ibid.

6. It can be shown that the broader the definition of the regulatory base (that is, the greater the value of inputs that are capitalized and are allowed in the base), the narrower is the gap at q^* between the RR profit-feasibility curve and the others. Indeed, when all inputs are capitalized for regulatory purposes, the RR path becomes identical with the MU path.

turn regulation nor with markup control *alone* is it possible to have the firm operate on inelastic portions of the average revenue curve. A firm will waste output or increase expenses through technological or economic inefficiency rather than sell its product at the negative marginal revenues associated with the price and output levels of inelastic average revenue schedules. This means that if the output q^* lies on an inelastic portion of the average revenue curve, the *MU* and *RR* profit-permissibility paths will not simply fail to reach a peak at q^*, they will not even include the coordinate q^*, since they stop earlier.[7] On the other hand, with a price-control strategy—profit-feasibility path *CP*—there are no such difficulties, even when average revenue is inelastic.

The difficulties for regulatory strategy associated with inelastic average revenue curves are not sufficiently appreciated in the literature on regulation. It should be realized, however, that an inelastic average revenue schedule does not imply inelastic demand and inelastic demand does not imply inelastic average revenue. The distinction is that the average revenue curve is often the sum of demand curves of markets that are spatially or temporally separate, or of market segments that are created through price discrimination, with regulatory authorities specifying the allowable price differentials. Movements along the average revenue schedule represent a weighted average movement along the demand schedules or along various parts of them, with weights determined by the regulatory process or by the costs of practicing price discrimination.

While the profit-permissibility schedule for price-control regulation is not bounded by the inelasticity of the average revenue schedule, it is bounded by a limit which, for lack of a better term, will be called the "competitive norm"—the output determined by the intersection of average revenue and *marginal* cost. That is, just as it is impossible with rate-of-return regulation or markup regulation to move the peak of the profit-feasibility curve to points where average revenue is inelastic, so it is impossible with a ceiling on the price level to move the profit peak to a range where marginal cost exceeds the ceiling price. This proposition is noteworthy only if average cost is a rising function of output; if average cost is a declining function of output, marginal cost is less than average cost, and such a range of outputs is dominated by the proposition stated earlier that regulation of monopoly cannot succeed if the profit-permissibility curve generates losses at its peak. But if average cost is a rising

7. Westfield, "Regulation and Conspiracy," p. 439.

function of output, marginal cost will be above average cost. And if the ceiling level of prices is pushed below the level of the competitive norm, the firm will want to reduce output by refusing load and creating shortages.

Throughout this discussion of the effects on permissible profit positions of alternative regulatory policies, it has been assumed that the regulatory rules are set so that each rule produces the same equilibrium price and output. It follows that for a given target *amount* of total profit allowed to the regulated firm, rate-of-return regulation would call forth less production than would the other techniques. Also, only slightly less obviously, if a given target rate of return is to be earned on capital, markup and price-control regulation would generate a larger output and a lower price.

This is regulation in theory; in practice, it is often a mixture of the three measures examined and of many other factors.

Innovations and Technological Change

There are many ways of categorizing innovations. From the point of view of the single regulated firm a distinction should be made between those that affect primarily the demand side and are reflected in the position of the firm's total revenue function and those that affect primarily the production side and are reflected in the position of its cost functions. Both affect the profit-possibility schedule and, more important from the standpoint of this chapter, the profit-permissibility schedules. Shifts in the cost function can come from shifts in the production function of the regulated firm or from changes in factor prices that are induced by shifts in the production functions of suppliers. Indeed, a given technological change that in one instance is transmitted through a market interface will, for a vertically integrated firm, lead to a shift in its production function.

Shifts in the revenue function can come from changes in the product of a regulated firm or the development of new uses for it. They are the result of shifts in the production and utility functions of the regulated firm's customers. One such innovation could even be improved transportation of the regulated firm's output to customers. While innovations from the cost side will raise the profit-possibility curve of the regulated firm, those from the demand side need not necessarily have the same effect. They might lower it. And whether either category of innovation raises or lowers the profit-permissibility curve depends on the mode of regulation.

The innovator for a given technological change can be the regulated firm, a firm that is its supplier, a firm that is its customer, or possibly another. Thus the economics of the innovative process in regulated industries, as elsewhere, may depend on whether the regulated firm is integrated backward into equipment production and forward into ownership of the appliances, as is the American telephone industry, or whether it is not, as in the case of the American electricity generating industry.

Finally, *pecuniary* innovations should be distinguished from *technological* innovations, a distinction reminiscent of that made about externalities by Jacob Viner in the early 1930s.[8] A pecuniary innovation is one that leads to a shift in the regulated firm's profit-possibility and profit-permissibility curves without shifting either its production function or any production function above or below it in the vertical production structure. Thus it is basically a marketing stratagem—such as creating a new category of consumers for whom price is lowered on a discriminatory basis in order to expand the market. The efforts of natural gas distributors to make inroads into electric energy markets (or vice versa) are examples of pecuniary innovations.[9] In contrast, the technological innovation shifts the technological production possibilities somewhere in the production structure.[10]

In the following analysis certain stylized categories of technological innovations that influence the production function of the regulated firm are considered. It is imagined that there are certain possibilities for intro-

8. *The Long View and the Short: Studies in Economic Theory and Policy* (Free Press, 1958), especially pp. 69–74.

9. Consider, for example, the "innovation" of "total energy systems," with on-site electric power generation using a gas fuel. It is basically a very old concept, which to be economic requires a high load factor (as, for example, in hospitals), a low price for gas, and a high price for electricity. But the going price of the centrally generated electricity may not correctly reflect the low cost of off-peak power, the low cost of fuel, and the substantial scale economies in the generation and distribution of electricity.

10. In practice it may sometimes be as hard to distinguish between these two types of innovations as it is to distinguish discriminatory from nondiscriminatory price differentials. For example, are unit coal trains primarily a technological or a pecuniary innovation? See Paul W. MacAvoy and James Sloss, *Regulation of Transport Innovation: The ICC and Unit Coal Trains to the East Coast* (Random House, 1967).

ducing changes in the production functions of the regulated firm, changes that might come either on the initiative of the regulated firm itself or on that of suppliers to the regulated firm. These possible technological changes compete with the status quo and will be promoted or adopted if there is an appropriate structure of incentives. Of course, if the market for an input is purely competitive, its supplier normally will have no incentive to promote anything at all. He can and does sell all he wants at the going price and can increase his equilibrium profits only by lowering his own costs. On the other hand, where markets for the inputs purchased are not purely competitive—the world of energetic salesmen, promotion, advertising, and labor unions—a supplier will have an incentive to generate demand for his product at the prevailing price. The latter is more likely to be the situation among suppliers to the regulated industries.

The costs of the technological changes themselves and of gathering the information needed for making them are not considered in this analysis; nor are changes in the prices of inputs purchased in response to technological change, and the important repercussions of these changes, evaluated. This chapter deals with some initial impacts of technological change, while neglecting some of its fuller general equilibrium aspects. And in the discussion of the effects of technological innovations on regulatory parameters, the regulatory agency is assumed not to respond to technological change by altering the regulatory technique or the magnitude of the regulatory parameter—the allowed markup, ceiling price, or rate of return, as the case may be. In this sense, the concept of regulatory lag is built into the analysis.

To further sharpen the focus of the analysis, a regulated firm that produces the output quantity q with only two inputs, L and K, is considered. The former (L) might be thought of as the services of labor, fuel, and variable inputs; the latter (K) as the services of machinery, equipment, and so forth, purchased by the firm. The analysis will be confined to stationary equilibrium situations, and no formal distinction will be made between the services of capital and amounts of capital. No specific theory of investment demand is implied by the models.

Before developing conclusions about the interrelation of technological change and regulation, equilibrium positions of regulated firms based on the profit-permissibility curves must be spelled out more fully.

Regulatory Equilibria

The production function for the regulated firm is

(1) $$q \leq F(K, L),$$

where the inputs K and L can be purchased at given prices r and w. The inequality emphasizes that a firm subject to regulatory constraint might find it profitable to be technologically inefficient. The production function is assumed to be subject to constant returns to scale, as is customary in the literature on the theory of technological change. The assumption is made here even though many economists might argue that there are usually increasing returns to scale in regulated industries. Actually there is probably no single category of production functions that would include all the production processes of regulated industries, and neither economies nor diseconomies of scale are assumed here.

Ceiling-Price Regulation

With ceiling-price regulation, the firm maximizes profit:

(2) $$\pi = R(q) - wL - rK,$$

where R is the total revenue function, subject to the regulatory constraint

(3) $$R \leq mq,$$

where m is the imposed ceiling price. Under the critical assumptions that $mq \geq wL + rK$ and that the price ceiling imposed by the regulatory commission is less than the monopoly equilibrium price, the equilibrium conditions are

(4a) $$q = F(K, L),$$

(4b) $$R = mq,$$

and

(4c) $$F_K/F_L = r/w.$$

That is, output q will be produced without technological waste; (4a) is an equality. The price charged, R/q, is exactly equal to the ceiling price m;

(4b) is an equality. And the equilibrium output is produced at lowest possible cost; the ratio of the marginal physical products of K and L (F_K/F_L) is equal to the ratio of their cost, r/w. Of course, these conditions imply that

$$(5) \qquad\qquad R'F_K < r, \quad \text{and} \quad R'F_L < w,$$

where R' stands for the slope of the total revenue function, because the price m is less than the monopoly equilibrium price. If the price ceiling were removed, the firm would generally use less L and K, raise its price, and reduce output so as to equate marginal revenue products with input costs.

Markup Regulation

With markup regulation, instead of expression (3) the regulatory constraint now is

$$(6) \qquad\qquad R - h(rK + wL) \leq 0,$$

where h is a percentage greater than unity, indicating the maximum markup on total costs that is allowed.

Under the further assumption that now $R' > 0$ and that again the regulatory constraint is effective, maximization of profit yields equilibrium conditions that are quite similar to those under ceiling-price regulation. In fact, conditions (4a) and (4c) hold as before, and (4b) is replaced by

$$(7) \qquad\qquad R - h(rK + wL) = 0.$$

This says that the firm will want to meet the regulatory constraint as an equality. In the process, it will produce output efficiently from a technical point of view (4a) and at least cost (4c), and reduce price and increase output relative to unrestricted monopoly, to the point where the markup is exactly equal to h (7). Thus inequalities (5) also continue to hold.

To simplify comparisons, it is assumed that the level of h is such that the equilibrium price, output, and therefore input configuration are the same initially as under the ceiling-price constraint. That is,

$$(8) \qquad\qquad R = h(rK + wL) = mq,$$

which, when solved for q, yields the magnitude q^* and the profit-permissibility maximum of Figures 2-2 and 2-3.

Rate-of-Return Regulation

If there is a ceiling on rates of return, profit is maximized subject to the regulatory constraint

$$(9) \qquad \frac{R - wL}{K} \leq \rho.$$

This leads to equilibrium conditions

$$(10a) \qquad q = F(K, L),$$
$$(10b) \qquad R - wL - \rho K = 0,$$
$$(10c) \qquad R_L = R'F_L = w,$$
$$(10d) \qquad R_K = R'F_K < r,$$

and

$$(10e) \qquad F_K / F_L < r/w,$$

where K is now identified as "capital," and ρ is the allowed ceiling rate of return. Here ρ is greater than r, so profits will be positive for some rate of production, and ρ is smaller than the rate of return associated with unconstrained profit maximization. In fact, for purposes of comparison, ρ is assumed to be set at such a level that the firm will choose to sell the output q^* at price m; regulation is thus effective in lowering monopoly price and raising monopoly output. If $R' > 0$, there is again no technical inefficiency (10a); the input-output configuration satisfies the regulatory constraint as equality (10b); and R_L, which is the marginal revenue of the marginal product of L—the input that is not included in the rate base—is equal to its given price, w (10c). In other words, the short-run marginal cost of producing output (varying L and holding K fixed) is equal to the marginal revenue R' of selling it. However, the marginal revenue of the marginal product of K—the input that enters the rate base—is less than its cost (10d). Thus, the output q^* is produced with a higher K/L ratio; the total cost of producing the output, as noted earlier, is higher than under the other two regulatory constraints; and total profit is less (10e). Also the rate of return to capital ρ is less than that under a price or markup ceiling. All these are corollaries of the Averch-Johnson effect.

How are these equilibria altered by potential technological changes? To answer this question, technological change must be characterized somewhat more specifically. First, the so-called Hicks-neutral changes in the regulated firm's production function will be considered, then the more general case of factor-augmenting technological changes. Again, it is not claimed that technological changes in regulated industries are likely to take such special forms; however, statisticians often have made such assumptions. By specifying a form, one can analyze it and then hope to study the sensitivity of the conclusions to changes in the specification.

Hicks-Neutral Technological Progress

Hicks-neutral technological change occurs when the production function for the regulated firm can be written

$$(11) \qquad\qquad F(A, K, L) = Af(K, L)$$

and the parameter A increases. That is, the technological progress takes place in such a way as to leave the isoquants unchanged in shape. Each new isoquant has a level of output associated with it that is a given percentage higher than the old isoquant in the same position.

In Table 2-1, the effect of alternative regulatory policies on the equilibrium of a firm is indicated. The columns headed $(dK/dA)(A/K)$ and $(dL/dA)(A/L)$ show, respectively, the percentage change in K and L consumed by the regulated firm in response to a 1 percent increase in the technological change parameter. The column headed $(d\pi/dA)(A/\pi)$ shows the percentage change in profits earned by the regulated firm in response to such innovations. The rows in the table indicate the various alternative regulatory techniques used. The results are obtained by differentiating the equilibrium conditions with respect to the technological change parameter and then making appropriate substitutions.

The symmetry of the effect on the demand for K and L under ceiling-price and markup control and the lack of symmetry under rate-of-return control is hardly surprising, since the technological change increases the productivity of both inputs proportionately.

The top row of Table 2-1 indicates that the demand for both inputs decreases in the same proportion as A increases. Thus under price control the equilibrium output of the regulated firm remains unchanged. The percentage increase in profits is equal to the ratio of total cost to total profits.

Table 2-1. Consequences of Hicks-Neutral Technological Change under Markup, Ceiling-Price, and Rate-of-Return Regulation

| Type of regulation | Adjustment elasticities of inputs | | Elasticity of profits $(d\pi/dA)(A/\pi)$ |
	K $(dK/dA)(A/K)$	L $(dL/dA)(A/L)$	
Ceiling-price	-1	-1	$\dfrac{wL + rK}{\pi}$
Markup[a]	$e - 1$	$e - 1$	$e - 1$
Rate-of-return[a]	$e - 1$	$\dfrac{-R_L}{R_{LL}L}\left[1 + \dfrac{R''q}{R'} + \dfrac{R_{LK}K}{R_L}(e-1)\right]$	$e - 1$

Note: For explanation, see the text.
a. e =the absolute value of the elasticity of the regulated firm's average revenue curve.

None of the benefits of the Hicks-neutral innovation will be passed on to customers. The innovation will lead to a "saving" of both inputs, so that sellers of inputs consequently have no incentive to promote such innovations among their customers. In fact, one can even imagine that suppliers of inputs would try to bring pressure to bear on the regulated firm or on the regulatory agency to prevent adoption of these technological changes.

The second line in Table 2-1 shows that the situation is quite different for markup regulation. Here e is the absolute value of the price elasticity of the average revenue curve, and it is of necessity greater than unity.[11] Instead of a decline in the demand for inputs, the identical technological change now will generate an increase in demand. As the total factor productivity increases because of technological change, the cost of producing each level of output can be reduced. Costs, therefore, will be reduced by the profit-maximizing firm, but the given percentage markup allowed and the elasticity of demand require that price be lowered and output increased so as again to satisfy the regulatory restraint on the relationship between revenues and costs. The percentage change in output is equal to e. The more elastic the demand, the greater is the proportionate effect on the output produced and inputs demanded. There are now also incentives for suppliers of K and L to promote such market-expanding technological changes. But these incentives do not come without some penalty. The increased demand, and presumably profit, for the suppliers arises because the increase in profits for the regulated firm is less than in the preceding

11. Otherwise marginal revenue R' would not be positive.

case.[12] In comparison to ceiling-price regulation, markup regulation increases the incentive to suppliers to promote technological change in the regulated firm, though at the cost of some decrease in the profit incentive to the regulated firm to adopt the technological change.

The third row in Table 2-1 shows that the effect of Hicks-neutral technological change on the rate-of-return-regulated firm is somewhat more complex and asymmetrical. Curiously, the percentage increase in the demand for K, the input entering the regulatory base, and in profits is the same as under markup control. It must be remembered, however, that the target output under rate-of-return regulation is produced with a higher K/L ratio because of the Averch-Johnson effect, so the absolute increase in the amount of K used is greater under rate-of-return regulation. Similarly, since the profit-permissibility curve has a lower peak under rate-of-return regulation, the absolute size of the profit increment from the same Hicks-neutral technological change is smaller than under markup regulation. An important unresolved issue relevant to regulatory policy is whether these effects of rate-of-return regulation do in fact give additional incentives to the suppliers of K to "sell" Hicks-neutral innovations, without seriously detracting from the incentive of the regulated firm to introduce them.

The effect on the demand for L is intricate. It depends in a complicated way on the average revenue and production functions. It can be positive or negative, greater or smaller than the corresponding term under markup control. In addition to the absolute value of the elasticity of the average revenue, e, three elasticities that determine these matters can be identified. The term outside the brackets is an elasticity of the marginal

12. This is because

$$\frac{wL + rK}{\pi} > e - 1,$$

which is evident since

$$e - 1 = R'q/(R - R'q).$$

Using Euler's theorem on homogeneous functions for the production function and substituting equation (4c),

$$R'q/(R - R'q) = R'F_L(wL + rK)/(Rw - R'F_L[wL + rK])$$
$$= \frac{wL + rK}{Rw/(R'F_L) - (wL + rK)}.$$

Since $w/(R'F_L) > 1$ because of condition (5), and $R - wL - rK = \pi$, the proposition is proved.

revenue product of L. It is positive because, as an equilibrium condition, R_{LL}—the slope of the marginal revenue product of L—must be negative. The expression $R''q/R'$ is the reciprocal of the elasticity of the firm's *marginal* revenue function. It is (algebraically) negative, provided that R'' —the slope of the marginal revenue curve—is negative. And $R_{LK}K/R_L$ is the reciprocal of a cross-elasticity of marginal revenue product—the proportionate change in K in response to a 1 percent change in the marginal revenue product of L. It is generally unrestricted in sign and magnitude. If by coincidence all three of these elasticities were to equal exactly unity,[13] the percentage response in the demand for L under rate-of-return regulation and markup regulation would coincide, although the absolute amount of the increase in L would be smaller because of the lower L/K ratio. But there are no straightforward relations among such elasticities and parameters of average revenue and production functions. The profit response is not affected by the magnitude or by the direction of the effect on the demand for L; for the regulated firm the changes in revenues from increases or decreases in amounts of L used are exactly offset by changes in costs.

To summarize, the gain in the regulated firm's total profits from a given Hicks-neutral technological change is largest, in both dollar and relative terms, under ceiling-price regulation. If the amount of additional profits to be earned by the regulated firm from new technology were the only factor governing its introduction, then ceiling-price regulation would provide the strongest incentive, although it does not encourage suppliers to develop new technology. On the contrary, they can be expected to use whatever political and economic pressure they can muster to prevent the introduction of the innovation. Benefits from the innovation will not accrue to suppliers, nor will they accrue to customers. All benefits become profits for the owners of the regulated firm. The percentage gains in the regulated firm's profits are the same under markup as under rate-of-return regulation, but in both cases they are lower than under ceiling-price regulation. The total gain in profits to the regulated firm is smaller under rate-of-return than under markup regulation. As has been suggested, if such profits were the critical incentive variable for technological change, both of these methods of regulation would call forth less technological change or a slower rate of growth. Whether markup or rate-of-return rules produce a greater innovative response depends on the extent to which the level of profits significantly affects the sensitivity to a potential change in total profits. If a

13. The sign-convention for elasticities used here is that own-elasticities are the absolute values and the cross-elasticity is the algebraic value.

businessman's Weber-Fechner law is operative, the potential proportional increase in profits determines the incentive to innovate.

Under both types of regulation, stimulus is also provided to a supplier of the input K. The Hicks-neutral technological change will, in both cases, generate an increase in demand for K that is proportional to the amount purchased. This turns out to be a larger quantity for rate-of-return regulation. If the *absolute* amount of demand increase that the supplier of K will experience determined his eagerness to effect technological change for his customers, and if the *relative* amount of profit to be earned by the regulated firm itself determined its eagerness for the technological change, then rate-of-return regulation would seem to have an edge on markup regulation in encouraging Hicks-neutral technological change. Under markup regulation, suppliers of the input L could similarly be expected to promote such innovation, while under rate-of-return regulation they might or might not, depending on the interaction of a host of technological and market considerations. In contrast to ceiling-price regulation, the benefits of technological change for both markup and rate-of-return regulation will to some extent be shifted forward in the form of lower price and increased output to the customers of the regulated firm and shifted backward in the form of added profit to its suppliers.

L- and *K*-Augmenting Technological Progress

It has been pointed out frequently in the literature that Hicks-neutral technological progress is a special case of factor-augmenting technological change, under the assumption of constant returns to scale. That is,

$$q \leq Af(K, L) = f(aK, bL),$$

where an increase in A (and a and b) leaves a/b constant.[14] A natural further step in the analysis is to turn to an examination of nonneutral Hicks technological progress, which is nevertheless factor-augmenting. A special case of this is Harrod-neutral technological progress, which is defined as a shift in the production function such that the average physical product of capital is constant provided the marginal physical productivity of capital remains unaltered. This is logically equivalent to b increasing while a is constant—a purely *L*-augmenting technological progress. This special case will first be examined, and then its mirror image—a purely *K*-augmenting technological progress—will be discussed.

14. An appropriate selection of units for K and L will allow $a = b$.

Table 2-2. Consequences of Harrod-Neutral (L-Augmenting) Technological Change under Markup, Ceiling-Price, and Rate-of-Return Regulation

Type of regulation	Adjustment elasticities of inputs		Elasticity of profits $(d\pi/db)(b/\pi)$
	K $(dK/db)(b/K)$	L $(dL/db)(b/L)$	
Ceiling-price[a]	$-\sigma\alpha$	$\sigma\beta - 1$	$\dfrac{wL}{\pi}$
Markup[a]	$-\sigma\alpha + e\alpha$	$\sigma\beta - 1 + e\alpha$	$\alpha(e-1)$
Rate-of-return[a]	$-\tilde{\alpha} + e\tilde{\alpha}$	$\dfrac{-R_L}{R_{LL}L}\left[1 + \dfrac{R_{LK}K}{R_L}\tilde{\alpha}(e-1)\right] - 1$	$\tilde{\alpha}(e-1)$

Note: For explanation, see the text.
a. e = the absolute value of the elasticity of the regulated firm's average-revenue curve
   σ = the value of the elasticity of substitution
   α = the output elasticity of L ($f_L L/q$)
   β = the output elasticity of K ($f_K K/q$)

The responses for the regulated firm to L-augmenting technological progress under alternative regulatory rules are summarized in Table 2-2. The first entry in the first row shows that under price-ceiling regulation Harrod-neutral technological progress, like Hicks-neutral technological progress, will generally cause a decline in the use of K. How great the decline is depends on characteristics of the production function—the magnitudes of the elasticity of substitution,[15] σ, and of the output elasticity of L, α, also called the "relative share" of labor.[16] In the limiting case of fixed coefficients of production, $\sigma = 0$, and K actually would not change. Under a Cobb-Douglas production function, $\sigma = 1$, and the output elasticity of L is the exponent α associated with the input L. The "easier" it is to substitute inputs (that is, the greater is σ), and the "more important" is L (that is, the greater is α), the sharper is the decline in the amount of K required.

The second entry in the first row shows that the demand for L will not necessarily fall in response to Harrod-neutral technological progress and ceiling-price regulation. This is in contrast to the effect of Hicks-neutral technological change in the corresponding entry in Table 2-1. If σ is large enough compared to β (the relative share of K), the demand for K will actually increase. Although the output of the regulated firm does not increase, the rise in efficiency, as it were, of input L will lead to an expan-

15. The elasticity of substitution is the percentage change in the proportion K/L in response to a 1 percent change in the marginal rate of substitution f_L/f_K along an isoquant. It is positive.

16. "Relative share" refers to the division of earnings under purely competitive conditions. Here it does not reflect earnings of the input.

sion of its use by replacing a large quantity of input K. Thus, even under ceiling-price regulation, it may be worthwhile for those who control input L to promote such market-expanding technological improvements.

The profit incentive for the ceiling-price regulated firm to introduce the Harrod-neutral innovation is given in the last column of the first row of Table 2-2. The magnitude of this entry is seen to depend only on the importance of the total cost of L purchased, compared to the regulated profit. The greater the ratio of these costs to profit, the greater is the percentage increase in profit from the Harrod-neutral technological change.

As the second line of Table 2-2 shows, the consequences of markup regulation are also somewhat different. While Hicks-neutral technological change generated an increase in the demand for both inputs under markup regulation, Harrod-neutral technological change does not necessarily lead to an increase in demand for both inputs. The magnitude of the elasticity of substitution relative to the elasticity of the average revenue function is critical. If $\sigma<e$, the substitution for K resulting from the L-augmenting technical improvement will be offset by the increased production resulting from the reduction in price that will be needed to avoid exceeding the allowed markup, since the technological change lowers costs. The greater e is in relation to σ and the greater α is, the larger the percentage increase in the demand for K will be. And with these larger increases in demand goes an assumed greater desire on the part of suppliers of K to promote such L-augmenting technological progress. On the other hand, if $\sigma>e$, demand for K must fall. And if L can so readily be substituted for K, producers of K can be expected to resist the introduction of these technological improvements.

The demand for L can also fall as a result of Harrod-neutral technological change and markup regulation. As Table 2-2 indicates, this will occur if the elasticity of substitution is sufficiently small and if α—the output elasticity of L—is small also. To give an extreme example, if $\sigma=0$ and α is less than $1/e$, the demand for L will fall. The increased output induced by the reduction in selling price that was necessary because of the fall in cost does not overcome the reduction in L made possible by the rise in its efficiency. Thus in such cases there will also be no incentive for suppliers of L to encourage these technological changes. It should be noted that in this situation there will be a substantial increase in the demand for K. Actually the effect on the demand for L can be viewed as depending on the weighted arithmetic average of σ and e, where the weights are the "relative shares" or output elasticities of K and L, respectively.[17] If this average is greater than one, the demand for L increases; if it is less than one, the demand decreases. The magnitude of the response depends on the difference between the average and one.

The profit response under markup regulation is, of course, positive also, but it is smaller than under ceiling-price regulation because a portion of the benefits are passed forward to customers as price is forced down along the elastic portion of the average revenue curve. By manipulating the relative profit increase for markup control in Table 2-2, one can obtain

$$\alpha(e-1) = \frac{wL + e(R_L - w)L}{\pi} < \frac{wL}{\pi}.$$

Thus the decrease in the profit response compared to that under ceiling-price regulation depends on the elasticity of the average revenue curve and also on the effectiveness of regulation as measured by the negative number $(R_L - w)L$.[18]

In the last row of Table 2-2 the effects of the Harrod-neutral innovation on the firm under rate-of-return regulation are summarized. In comparing these entries with the others in the table, it must again be remembered that the K/L ratio for the given output q is larger here and maximum profit is lower because of the Averch-Johnson effect. Also the functions f_L, R_L, and so forth, will take on different values, since their arguments take on different values. Unfortunately, the notation used does not emphasize these differences sufficiently, although the tilde over some of the symbols should serve as a reminder.

What can be said about $(dK/db)(b/K)$ under rate-of-return regulation? Consider first the case where $\sigma = 1$. As under Hicks-neutral technological change, the percentage change in the demand for K is exactly the same under rate-of-return as under markup control. Here, it is important to remember that α does not vary with increases in the K/L ratio when $\sigma = 1$, for this is the Cobb-Douglas case—the case of constant output elasticities. Hicks-neutral and Harrod-neutral technological changes cannot be distinguished in the Cobb-Douglas production function. And in assessing the incentive to suppliers of K, it should be noted again that the absolute mag-

17. Note that $\alpha + \beta = 1$ because of the linear homogeneous production function.

18. $(R_L - w)L$ normally becomes larger in absolute value as regulation pushes the firm to produce output at the peak of the profit-permissibility curve—an output in excess of the monopoly equilibrium and the peak of the profit-possibility curve, where $R_L - w = 0$.

nitude of the increase in demand for K is larger under rate-of-return regulation than under markup control. Suppose, on the other hand, that $\sigma > 1$. Then, if α could be constant as before, markup regulation would generate in this case of L-augmenting technological change a smaller percentage change in demand for K than would rate-of-return regulation. But α is not constant when $\sigma > 1$ and the K/L ratio varies. It is easily shown that if $\sigma > 1$, then α is smaller the higher the ratio K/L. Similarly, if $\sigma < 1$, then α is larger the higher the ratio K/L. So the percentage effects on the demand for K if $\sigma > 1$—moving from the markup control equilibrium to the rate-of-return control equilibrium—depends on whether the ratio $\tilde{\alpha}/\alpha$ of output elasticities is greater or smaller than the ratio $(-\sigma + e)/(-1 + e)$. Evidently, for an elasticity of substitution as high as $\sigma = e > 1$, rate-of-return regulation leads to a greater percentage, as well as absolute, increase in the demand for K than does markup control.

Skipping over the $(dL/db)(b/L)$ entry for rate-of-return regulation for the moment, one can see that the percentage increase in profit under rate-of-return regulation compared to that under markup regulation also depends on the elasticity of substitution. If $\sigma = 1$, the output elasticity for L is the same in both cases—that is, $\alpha = \tilde{\alpha}$—so the percentage increase in profit is the same. But it must be remembered that the absolute increase in profit is lower for rate-of-return regulation. If $\sigma < 1$, however, then $\tilde{\alpha}$, the output elasticity for L associated with the higher K/L ratio for rate-of-return regulation, is larger than α, and the profit response in percentage terms is greater; on the other hand if $\sigma > 1$, the converse holds. Thus, no simple generalization is possible about relative profit incentives for the regulated firm under markup and under rate-of-return regulation. Which regulation provides a greater encouragement to technological change depends, for a Harrod-neutral innovation, not only on whether percentages or absolute magnitudes are relevant, but also on a technological parameter, that is, σ.

The responsiveness of L to Harrod-neutral technological change and the push for innovation to be expected from its suppliers is less transparent. In Table 2-2, the term outside the bracket is the elasticity of the marginal revenue product of L function; $R_{LK}K/R_L$ is the reciprocal of a cross-elasticity giving the percentage change in the marginal revenue product of L in response to an incremental 1 percent change in the rate of use of K; $\tilde{\alpha}$ is the output elasticity of L (its "relative share"); and e is the familiar absolute value of the elasticity of the average revenue schedule for q. Since the cross-elasticity term can be either positive or negative, depending on both market demand and production conditions, $(dL/db)(b/L)$ can be either positive or negative. A comparison with the same expression under markup regulation is also inconclusive. Some additional insight may be obtained by inquiring how the change in demand for L in response to an L-augmenting technological change under rate-of-return regulation is related to the change in demand in response to a change in the price of L in the same regulatory environment. On substitution, one obtains

$$\frac{dL}{db} \cdot \frac{b}{L} = -\left(1 + \frac{dL}{dw} \cdot \frac{w}{L}\right),$$

where $(dL/dw)(w/L)$ is the algebraic value of the price elasticity of the firm's demand for L, which is, of course, negative.[19] If L is supplied to the regulated firm by a monopolist who has tailored the price w to this particular customer, then $(dL/dw)(w/L) < -1$, because a monopolist sells only on an elastic portion of his demand curve. Consequently, $(dL/db)(b/L)$ would have to be positive and thus give this monopoly seller of L an incentive to foster L-augmenting technological change in the production process of the regulated firm in order to stimulate his own demand.

In Table 2-3 the effects of K-augmenting technological progress are summarized. Because of symmetry, the effects under ceiling-price control and markup control can be obtained immediately by inspecting Table 2-2; they require no further discussion. For rate-of-return regulation, however, the augmented input is now the one entering the rate base. The surprising result here is that the effect on the demand for K and the amount of profit earned also show the same symmetry. Thus, it can be deduced at once that under rate-of-return regulation the demand for the augmented input K must increase, since $\tilde{\beta}$ is positive and $e > 1$. It will be remembered that with L-augmenting technological progress the demand for the augmented input may or may not rise.

How does the size of the demand response for K under rate-of-return control compare with that under markup control? The response in the demand for K with K-augmenting technological change under markup control is

$$\sigma\alpha - 1 + e\beta = (\sigma - 1) - \beta(\sigma - e).$$

When $\sigma = 1$, the percentage (but not the absolute) effects under markup and rate-of-return control are once again the same. If, however, $\sigma = e > 1$,

19. This follows from the stability conditions associated with the maximum under constraint.

Table 2-3. Consequences of K-Augmenting Technological Change under Markup, Ceiling-Price, and Rate-of-Return Regulation

Type of regulation	Adjustment elasticities of inputs		Elasticity of profits $(d\pi/da)(a/\pi)$
	K $(dK/da)(a/K)$	L $(dL/da)(a/L)$	
Ceiling-price[a]	$\sigma\alpha - 1$	$-\sigma\beta$	$\dfrac{rK}{\pi}$
Markup[a]	$\sigma\alpha - 1 + e\beta$	$-\sigma\beta + e\beta$	$\beta(e-1)$
Rate-of-return[a]	$-\tilde{\beta} + e\tilde{\beta}$	$\dfrac{KR_{LK}}{LR_{LL}}\left[1 + \tilde{\beta}(e-1)\right]$	$\tilde{\beta}(e-1)$

Note: For explanation, see the text.
a. e = the absolute value of the elasticity of the regulated firm's average revenue curve
   σ = the value of the elasticity of substitution
   α = the output elasticity of L ($f_L L/q$)
   β = the output elasticity of K ($f_K K/q$)

the percentage increase in demand is greater under markup control than under rate-of-return control. This occurs even though $\tilde{\beta}$ is now larger than β because of the higher K/L ratio of rate-of-return regulation. On the other hand, for values of σ near zero the demand response for K under markup regulation becomes smaller and can even be negative. Rate-of-return regulation, unlike markup regulation, must always encourage the promotion of K-augmenting technical progress by suppliers of K, as implied by the Averch-Johnson effect.

The profit incentive for the rate-of-return-regulated firm of K-augmenting innovations is greater or smaller in percentage than that for the markup-regulated firm, depending on whether σ is greater or smaller than unity. As was suggested above, the higher K/L ratio for rate-of-return regulation leads to $\tilde{\beta} > \beta$ when $\sigma > 1$, and $\tilde{\beta} < \beta$ when $\sigma < 1$.

The response of L, here the nonaugmented factor and the factor not entering the rate base, is again ambiguous. It will be positive or negative depending on the sign of R_{LK}, a substitution-complementarity measure between K and L. Its sign depends both on the slope of the marginal revenue curve for output sold and on the properties of the production function. However, since R_{LL} is negative, complementarity as measured by a positive R_{LK} would lead to a decrease in demand for L; substitutability would lead to an increase.

It is difficult to summarize the consequences of factor-augmenting technological progress. Much depends on parameters of the production function and the average revenue function. In all cases considered, the regulatory formulas provide some profit incentive for innovation by the regulated firm. As was the case for Hicks-neutral technological progress, potential profits of regulated firms increase. The largest percentage and absolute increases in profits for the regulated firm occur under ceiling-price regulation. This is true for L-augmenting and K-augmenting progress, and it was seen to be true also for Hicks-neutral progress. The percentage effect under markup regulation, however, is no longer the same as the percentage effect under rate-of-return regulation. Which is larger depends on the relationship between α and $\bar{\alpha}$, or between β and $\tilde{\beta}$, and this is related to the magnitude of σ. If $\sigma > 1$, then $\tilde{\beta} > \beta$, and $\bar{\alpha} < \alpha$. So, in this case of "easy" substitution, L-augmenting technological progress gives a higher percentage increase in profits under markup regulation, while K-augmenting progress gives a higher percentage increase in profits under rate-of-return regulation. But if $\sigma < 1$, then $\tilde{\beta} < \beta$ and $\bar{\alpha} > \alpha$, and the conclusions are exactly reversed. In contrast to Hicks-neutral technological change, the effects of factor-augmenting technological change on demands for factors are no longer clear-cut. Easy substitution possibilities may lead to increases in demand for the augmented factor under ceiling-price regulation at the same time that there are relatively sharp decreases in the demand for the other factor. Somewhat more difficult substitution between factors will lead to the same consequences as Hicks-neutral technological change and ceiling-price regulation: a decrease in demand for both factors. When $\sigma < 1$, the augmented factor has a sharper percentage cutback in demand than does the nonaugmented factor. The arrangement of Tables 2-1, 2-2, and 2-3 emphasizes the fact that the effects of markup regulation on factor demands can be broken down into the effect of a ceiling price by itself, with an output effect (whose magnitude depends on the elasticity of the average revenue function and on the relative share or output elasticity of the augmented factor) added on as price is reduced and production increased so as not to exceed the permitted markup on cost. Thus under markup regulation there is always a greater positive, or smaller negative, impact on demand for factors than there is under ceiling-price regulation. The resulting stimulus for the innovation is therefore algebraically greater, the more important the augmented factor is in the regulated firm's production function (that is, the higher its relative share) and the greater the elasticity of demand for the regulated product. The consequences of rate-of-return regulation in relative terms for the factor K entering the rate base were seen to depend on σ as well as on relative shares. The magnitude of the effect is the relative share of the augmented factor multiplied by the excess above unity of the elasticity of the average revenue curve. Surprisingly, whether the augmented factor enters the rate base or not is of no importance for the demand stimulus. For the factor L, relative demand responses are not readily described in terms of the parameters considered here. Other parameters involving derivatives of the marginal revenue product curve become important.

Demand-Creating Innovation

The regulated firm may succeed in shifting its profit-possibility and -permissibility schedules by promoting technological changes in the production processes of its customers, just as do its own suppliers. The above analysis suggests some of the circumstances under which a regulated firm may have an incentive to encourage innovation by customers who are themselves subject to regulation. One need merely note that output q of the regulated firm may also be an input like K or L in the production functions of regulated customers of the firm. In situations like those in Tables 2-1, 2-2, and 2-3, where demand responses are positive, the regulated firm will want to devote some of its promotional-innovational efforts to shifting its own profit-permissibility curve upward. Where demand responses are negative, the regulated firm would be expected to do the reverse. Even for the special, stylized forms of technological change considered for customers' production functions, the effects of the change on sales generally depend on technological and market parameters as well as on the regulatory environment. Where customers are regulated so that *they* are required to pass forward in the form of lower prices some of the gains derived from technological changes, as under markup regulation, the regulated firm will generally have a greater incentive to encourage the adoption of new technology than where customers are not required to do so, as under ceiling-price regulation.

If customers are unregulated firms, the circumstances under which the regulated firm will or will not benefit from the technological changes it induces among them will surely also depend on the structure of the markets in which the customers must sell their products. Since this chapter is concerned with the regulated sector in the vertical production structure, these problems are not considered further, although they are obviously important.

Structure of Input Markets

The preceding analysis shows that technological change generally leads to shifts in demand for inputs. These demand changes provide stimulus for or against the promotion of technological change by the producers of the inputs. Moreover, the magnitude of the effect on the profits of the regulated firm was considered to be a measure of the intensity of the incentive for the firm to undertake an innovation. It must be emphasized again that the secondary repercussions, both on profits of the regulated firm and on suppliers' incentives, from input price changes resulting from technological changes have not been taken into account here, nor have any changes the regulatory commissions may make in the magnitudes of the regulatory parameters. Generally, some changes in prices of inputs might be expected when demand for them changes or when suppliers seek through technological change to promote the use of inputs they provide. The magnitude of these price repercussions, if there are any, is determined by the structure and behavior of the input markets. A wide range of price changes is theoretically possible, depending on the nature of the competition, on the elasticity of supply if the market is competitive, and on many other factors if the market is oligopolistic and monopolistic.

The analysis should be extended to include such price repercussions on the input side, and indeed it should be carried still another step backward on the supply side and another step forward on the demand side. Similarly, if one had better theories on the behavior of regulatory commissions, one would want to study the effects of technological changes on the magnitudes of the regulatory parameters. This chapter offers only a partial equilibrium analysis of problems that have general equilibrium feedback.

Two propositions about the effect of the structure of input markets can be suggested. First, if input markets are almost purely competitive, the incentive for suppliers to promote innovations in the production functions of their customers is very low. The reason for this is found in the writings of Edward H. Chamberlin[20] and Joseph A. Schumpeter:[21] the competitive seller cannot hope to reap the rewards of his own technical promotional

20. *The Theory of Monopolistic Competition: A Re-orientation of the Theory of Value* (Harvard University Press, 1948).

21. *The Theory of Economic Development: An Inquiry into Profits, Capital, Credit, Interest, and the Business Cycle* (Harvard University Press, 1934).

efforts. Second, under monopoly control of an input, the supplier will try to create a scarcity in technological information—that is, on the technological parameters valuable to his customers—for the same reason that one would expect him to contrive a scarcity of the input he sells. Thus a telephone system or an airline that buys equipment from monopolists, rather than being its own vertically integrated supplier, might for these reasons develop more slowly technologically. Many economists would argue a priori the reverse of this last proposition.

Conclusions

In this chapter some aspects of the interrelation of innovative behavior and methods of regulation have been explored. The analysis presumes, on the one hand, clearly defined possibilities for innovation and, on the other, specific regulatory rules laid down by a regulatory commission. A firm is assumed to act so as to maximize its profits, and the regulatory commission is assumed not to respond to the consequences of the innovation by changing the rules.

The analysis here needs to be extended in many directions. Some of these, relating to input price changes caused by innovative behavior, have already been referred to. Others relate to dynamic effects. How does technological change influence the growth rate of the firm under alternative regulatory constraints? How does regulatory policy influence the durability of capital, and therefore replacement needs, and therefore the rate at which new technology embodied in inputs will be absorbed by the regulated firm? Answers to some of these questions will require much more complicated models of the regulated firm. Other desirable extensions of the analysis would be based on a more sophisticated theory of the behavior of regulatory commissions. Unfortunately, such theory quickly encounters the indeterminacies associated with bilateral monopoly and bargaining theory.

It must be emphasized once again that the limited yet rigorous findings presented here are based on the postulate that the market opportunities of the regulated firm generate positive marginal revenue—that the average revenue curve has a price elasticity greater than unity. When this is not the case—that is, where average revenue is (incipiently) inelastic—the only one of the three regulatory instruments that can help enforce a discipline of technical and economic efficiency in a profit-motivated firm is price-control regulation. All of the conceivable indirect methods of control based on cost do not even encourage the regulated firm to economize. Changes in profits of the regulated firm as a result of technological changes will not be positive. A regulatory commission may help to ensure that the average revenue curve is elastic by specifying that price changes in different markets, or among different products, bear a predetermined relation to each other, so that, for example, a price reduction in one market requires reductions in others in amounts that lead to an *increase* in revenue. But if all conceivable price decreases lead to *decreases* in total revenue, average revenue is surely *inelastic*, and indirect methods of control, such as rate-of-return or markup regulation, cannot even promote operation *on* the production function. They encourage technological and economic waste. Obviously, economic theory cannot be used to predict innovative behavior under circumstances where incentives to economize are completely lacking.

CHAPTER THREE

Scale Frontiers in Electric Power

William R. Hughes

THE EFFICIENCY OF the electric power industry is strongly influenced by the interaction of scale-related technological change, industry structure, and regulation. This chapter is an outgrowth of research into the efficiency aspects of the organization of the power industry that relate closely to the adoption and diffusion of new technology.[1]

The indefinite zone where efficiency and progress meet and interact is important to utility performance and to the role of electric utilities in the process of technological change. Advances in electric power technology have taken the form of a steady cumulation of many individually minor changes. All are closely linked, within a common technological setting, to one another and to the economics of power supply in relation to changing demand.[2] In general, productivity has advanced rapidly, far faster than the average for the entire economy.[3] Both technological advances and, especially in distribution, gains in technical efficiency from realizing well-established scale economies have contributed to this performance.

Electric utilities have no corporate links with equipment manufactur-

1. Resources for the Future, Inc. (Washington), is sponsoring the efficiency research. The results are reported in a paper as yet unpublished.
2. Philip Sporn, one of the most active promoters of technological progress in this industry over the past several decades, presents an interesting interpretation of technological change in power supply in *Technology, Engineering, and Economics* (MIT Press, 1969).
3. According to Kendrick's measurements, the long-term rate of increase in total factor productivity in electric power has been the highest of any U.S. industry—more than three times the economy-wide average. John W. Kendrick, *Productivity Trends in the United States* (Princeton University Press for National Bureau of Economic Research, 1961), pp. 136–37.

44

ers. The latter carry the main burden and capture the most direct commercial rewards of private research and development in electric power technology. Because of this division, an explicit regulatory policy,[4] the principal role of electric utilities in technological change has been to encourage new developments. Their willingness to undertake pilot projects or to adopt technology that is not fully developed—sometimes with special price inducements from manufacturers—affects the rate at which experience accumulates and new developments diffuse. The utilities' alertness in perceiving opportunities and aggressiveness in pressuring manufacturers to develop improvements are important influences on the course of technology.

Economists from Adam Smith to Robert M. Solow have recognized that economies of scale and technological progress are often closely linked. The realization of latent scale economies is an especially important form of technological progress in the utility industries. This has significant consequences for regulatory policy. For the institution of regulated monopoly to work in the public interest, these economies must persist over long periods of time while technological change and economic growth continue. The same characteristics of dynamic production that dictate monopoly in local distribution govern the economies of spatial integration in the generating and transmission, or "bulk-power," stages of electric power supply.

Fundamental questions must be raised about the relations among regulatory policy, power industry organization, and the industry's economic performance. Since the mid-1960s, the industry has been undergoing its first substantial consolidation since the holding-company era of the 1920s. The principal stated purpose is to exploit economies of scale.

How far should the merger movement be allowed to go? Do current regulatory policies guarantee that the industry will end up with a high-performance structure? What, if any, performance effects can be predicted as the industry becomes increasingly concentrated? Both the direction of these effects and their magnitude at successive concentration levels are important.

The evidence reviewed in this chapter suggests that both technical effi-

4. The breakup of holding companies under the Public Utility Act of 1935 severed all ownership ties with manufacturing and engineering firms in order to eliminate the widespread practice of "milking" captive utilities with excessive prices that were ultimately passed on to ratepayers.

ciency and scale-related technological progress have been closely tied to the organization of the industry.

Bulk-Power Economics and the Coordination Problem

Two stages of bulk-power supply are production and transmission of electrical energy. The basic production unit is a network that links generating stations and bulk-power distribution points in a single inter-connected complex that often covers a very large geographical area. A third stage—distribution—is inherently local in nature; an efficient distribution network need not be very large and its management can easily be separate from that of its bulk-power source.

The geographical extent of an efficient bulk-power network, as well as the amount of internal interconnection, is determined by the balance between the benefits and the costs of interconnection. This balance has tended to favor increasing interconnection over time as demand has grown and technological progress has reduced transmission costs. The present large U.S. networks were formed by a long process of amalgamation of what are now subnetworks. The United States is already close to being a two-network country, and the process of interconnection across the Rockies to link the two networks has already begun. The larger network, which has over 75 percent of U.S. generating capacity, embraces several hundred generating plants and a considerably larger number of distribution points. The planning and operation of these large networks is in the hands of about 1,300 separately managed "systems," 100 of which were large enough to account for 89 percent of the power generated in 1962.[5]

Organizational Implications of Large Networks

Because these large networks are organized on a multisystem basis, coordination is a problem. Each system controls only a part of the network with which it is, or should be, interconnected. "Coordination" is a catchall term that embraces all forms of cooperative economizing by two or more systems. It is a means of reducing or eliminating the gap in effi-

5. *National Power Survey,* A Report by the Federal Power Commission, 1964 (Government Printing Office, 1964), Pt. 1, p. 17.

ciency between the network optimum level and the level that would result if systems were separately planned and operated.

The central fact of bulk-power economics is the interdependent, indivisible nature of the whole network. A high degree of technical coordination in design of facilities and in operating procedure is needed if the network is to function at all. Furthermore, the economic interdependence of the network's component parts has far-reaching implications for optimum scale, design, location, and operation of its facilities.

The standard of economic efficiency is the optimum for the network as a whole. This statement of hypothetical perfection, however, is of little practical value, especially when it is applied to networks covering large geographical areas. A one-firm industry would find it necessary to make most of its planning and operating decisions at subnetwork levels. The most important differences in cost between uncoordinated performance and optimum performance arise from network economies that are exploited largely by optimization over geographic areas much smaller in scope than the entire network. Consequently, the principal advantages of an efficient coordination pattern can be achieved within subnetworks that are of an appropriate size; further coordination is largely a matter of providing interconnections among subareas and of reconciling decisions at the subnetwork level to make use of special opportunities for joint economy.

Economies of Coordination

In view of the extreme capital intensiveness of the industry and of the important effect of planning decisions in restricting operating choices, it is not surprising that coordinated planning of new capacity accounts for most of the cost effect of coordination. The most important role of operating coordination, especially in load-frequency control, emergency procedures, and maintenance scheduling, is in making the coordinated network feasible and reliable. If each system were to optimize in isolation, the costs connected with planning decisions would probably exceed the network optimum by more than $1 billion a year. In contrast, the analogous cost gap in operating decisions would probably be measured in tens of millions of dollars. According to calculations made for the Resources for the Future study, actual coordination has reduced the potential cost gap by more than half.

Overall coordination leads to economies of scale in equipment components and to the pooling of reserve capacity; these are closely related and are best considered together. If increased coordination were artificially limited to the achievement of potential economies from these two sources alone, the cost gap between existing and potential coordination would be cut by more than half.[6] Other economies—chiefly in load diversity, location, and design of facilities—are exploited largely at levels of geographical integration within those required for exploiting economies of scale and reserves. Moreover, the various benefits of coordination tend to reinforce each other. A network whose economies of scale and reserves are well coordinated is likely to be well coordinated in other important respects.

Efficient Scale for Generating Equipment

Both economies of scale in equipment and in pooling of reserve capacity apply to transmission as well as generating facilities, on which most of the evidence in this chapter is concentrated. Generating costs amount to some 80 percent of total bulk-power costs,[7] and the achievement of scale and reserve economies in generating tends to be complementary to the achievement of similar economies at the transmission stage.

Steam-electric capacity is added in increments of generating units, consisting of integrated boiler-turbine-generator combinations. Substantial and consistent economies of scale persist up to the largest sizes of units that have been built. Additional, though less substantial, economies of scale exist at the plant level for multiunit stations. The elasticity of total generating cost with respect to unit size is variable from case to case. It averaged around 0.8 for scales appropriate to the 1950s but rose as unit sizes increased and was perhaps 0.9 for the large conventional units added in the 1960s.[8]

Within the range of capacity in which substantial operating experience has been gained, efficient unit sizes for the network represent a balancing

6. The quantitative and analytical bases for the efficiency gap statements are developed at length in the author's Resources for the Future (RFF) study.

7. See the author's RFF study for documentation of this and other cost data used in this chapter.

8. These figures are representative of the consensus of engineering sources and receive some corroboration from relevant econometric studies.

of incremental scale economies in generation against the aggregate effect of several network penalties: reserve capacity effects, transmission cost effects, and the cost of temporary overcapacity as a result of building in advance of demand. Under load conditions relevant to three-quarters or more of total U.S. thermal capacity additions, it appears that a major limiting factor to *efficient* unit size has been the frontier of technically feasible unit sizes.

The Nature of Scale Frontiers

The scale frontier is a range rather than a specific limit. Within this range, costs become increasingly uncertain as scale increases, and there is great scope for variation in managerial judgment as to the appropriate scale. Transmission costs and reserves may have more importance within the frontier zone than at established unit sizes because of the flattening of the effective average cost curve within the frontier zone, reflecting the shakedown costs and risks associated with pioneering units.

The relations embodied in the scale frontier idea are shown in Figure 3-1. The vertical portions of the C_{CH} and C_{CL} curves indicate the technical impossibility at a given time of building units that are larger than some critical scale. The technical problems to be overcome in extending the frontier may be major or minor, depending on the principal operative bottlenecks. The development of pioneering units involves mostly high-level design work. The cumulative progress involved in extending the frontier over long periods of time is frequently very substantial. In producing the necessary technology during the period from year t to, say, $t + 10$, use may be made of the results, unavailable at t but forthcoming in the interim, of research and development by equipment manufacturers as well as other sources of progress in such fields as metallurgy and combustion.

The weight of empirical evidence, including econometric studies, and the planning assumptions used by engineers in the industry strongly suggest that postwar technological progress in conventional steam-electric generation has been closely associated with the extension of scale frontiers. In Figure 3-1, shifts in the cost function take the form of extending the C_T and C_C curves into new territory, shown schemati-

Figure 3-1. Cost Function for New Base Load Unit[a]

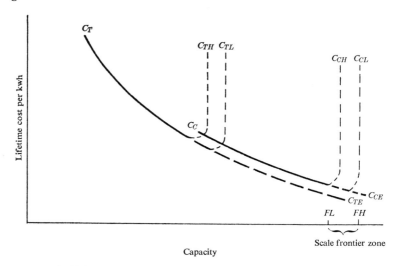

C_T =(solid line) cost curve for established scale with tandem compound design
C_C =(solid line) cost curve for established scale with cross-compound design
C_{TH} (or C_{CH}) =high-range estimates of costs for tandem (or cross-compound) design within scale frontier zone
C_{TL} (or C_{CL}) =low-range estimates of costs for tandem (or cross-compound) design within scale frontier zone
C_{TE} and C_{CE} =continuation of the C_T and C_C curves expected once technology has expanded the established scale range
FL, FH =lower and upper bounds to the frontier zone

a. The drawing abstracts from intertemporal shifts in the generating cost function for scales within previously established frontiers.

cally by C_{TE} and C_{CE}. Downward shifts in the cost function for already established unit sizes have apparently been of minor importance.[9] Parallel scale frontier phenomena can also be observed in transmission, where scale economies are especially pronounced, with cost elasticities of 0.5–0.6 for major transmission components.

Though the scale frontier operates largely as an externally imposed constraint on the decisions of the individual utility, the frontier is also

9. The principal relevant econometric studies are Ryutaro Komiya, "Technological Progress and the Production Function in the United States Steam Power Industry," *Review of Economics and Statistics,* Vol. 44 (May 1962), pp. 156–66, and Yoram Barzel, "The Production Function and Technical Change in the Steam-Power Industry," *Journal of Political Economy,* Vol. 72 (April 1964), pp. 133–50. L. K. Kirchmayer and others, in "An Investigation of the Economic Size of Steam-Electric Generating Units" (*AIEE Transactions,* Vol. 74, Pt. 3 [1955], pp. 600–09), provide a typical example of the shift assumptions used in system planning exercises.

demand-related for the industry as a whole. Manufacturers concentrate their development efforts on generating equipment their customers are likely to want, and development work on future scales tends to be synchronized with the expected demands of a growing industry.[10] Aggressiveness on the part of utilities in requesting new developments contributes to the process.[11] Moreover, the rate at which experience is acquired at each new size affects the rate of advance of the scale frontier. Each new unit within the frontier zone represents a step in the "proving out" process by which a new scale becomes established. Similarly, each frontier unit adds to the necessary foundation of experience for further extensions of the scale frontier.

Scale Frontiers and Industry Performance

The scale-frontier phenomenon raises interesting questions about the potential impact of regulation on performance, especially in the areas of merger policy and regulatory influences on coordination. In particular, how rapidly should scale frontiers advance in a high-performance power industry? How does current performance compare with the potential? What regulatory approaches would improve the performance of industry in advancing scale frontiers?

The Optimal Rate of Introduction

Theoretically, the economically optimal rate of advance in scale frontiers should synchronize development by manufacturers and adoption by

10. Published studies of the choice of unit sizes, sponsored by manufacturers, utilities, and engineering firms, usually phrase their assumptions concerning unit sizes available for future additions in these terms. See, for example, Kirchmayer and others, "Steam-Electric Generating Units."

11. Forrest McDonald gives a classic example in his biography of Samuel Insull. Insull was not an engineer but had studied political economy at Oxford and was an admirer of the work of Adam Smith. He played a major role in forcing the introduction of commercial-scale steam turbines at the turn of the century, when the scale limits of reciprocating steam engines had been reached. Overriding the claims of experts at General Electric and his own engineering staff that steam turbines were not feasible, Insull obtained GE compliance by threatening to go to foreign manufacturers. Thereafter, Insull continued to demand large extrapolations of scale which gave a strong assist to the major expansion of unit sizes that occurred from 1900 to 1930. *Insull* (University of Chicago Press, 1962), pp. 98–101.

utilities so as to achieve a marginal balance between prospective benefits and costs of increased frontier-advancing activity. Regulation can influence the degree to which this balance is achieved, primarily by affecting the demand (or benefit) side of the balance. Benefits are measured in expected cost reductions to the industry resulting from an advance in the scale frontier. These benefits are registered in the market for electrical equipment through the size mix of prospective orders for new units. If institutional factors produce too many inefficiently small units, the scale frontier will itself be suboptimal, and the rate at which new scale-related technology is adopted by the industry may be too slow.

Even under the most favorable conditions for advancing the scale frontier, the cost side of the equation imposes fairly strict upper limits on the economical pace of advance, and trying to force the pace could mean sharply rising cost of development. The experience required for pushing out the scale frontier is related to time and cannot be acquired by increasing the number of similar new units. Perhaps the greatest uncertainties connected with units arise from problems that may not show up until the units have been in operation a few years. For the industry as a whole, the socially optimal number of pioneering units during the first two or three years of any major advance in scale, design, or steam conditions is probably rather small, most often ranging from perhaps two or three to half a dozen. The only structural requirement for achieving an optimal pace of introduction is that enough systems be able and willing to adopt new technology at an appropriate rate.

Although it is not feasible to indicate a specific optimal rate of progress in scale frontiers, the elements of cost and demand set broad limiting values against which the actual record can be compared. On the demand side, the association of industry organization, regulation, and the size mix of new generating units can be examined. Absolute rates of introduction give some gross evidence of the degree to which this pace could be accelerated before time-related cost penalties are likely to be severe. If large increases in the frontier follow one another at short intervals and each new frontier shows only a few years between initial pioneering and production in fairly large amounts, development would probably not have advanced much more rapidly under the most encouraging circumstances.

In some cases developments may have been adopted prematurely. A high-performance industry inevitably makes mistakes in both directions, whereas a symptom of too slow an advance in scale frontiers may be the infrequency of apparent mistakes in the opposite direction. On the other hand, frequent and costly bandwagon movements into unproven territory suggest that the industry is reacting to forces that are pushing the new developments too rapidly. These forces might include technological rivalries combined with irrational or status-oriented patterns of imitation, and the effect of the rate-regulation system in underwriting technological risks, thereby possibly encouraging overexpansion to enlarge the rate base.

Industry Organization and Scale-Related Performance

Advancement of scale frontiers at a socially optimal pace, as well as an efficient mix of unit sizes within established frontiers, requires a combination of industry organization and coordination that produces planning decisions consistent with network optimality. The organization and performance of the industry from this standpoint should be reviewed.

The large U.S. networks are planned and operated in a complex process that includes a mixture of independent and coordinated decision making by a large number of managerial units or systems: holding companies, independent operating companies, various public power entities, and cooperatives. Primary interest here is in the privately owned, regulated sector, which has approximately 80 percent of the total generating capacity of the industry and over 90 percent of the thermal generating capacity. Both the private sector and the industry as a whole rely predominantly on steam-electric generation, and the share of thermal sources is increasing.

An examination of the size distribution of bulk-power systems as it relates to efficient unit size shows the amount of coordination or other adaptation that is needed to bridge the gap between independent decision making and the overall optimum. Tables 3-1 and 3-2 give summary information on this aspect of industry organization, dealing with the total industry and the private sector, respectively.

An electric utility system that is too small to justify capacity additions of efficient scale by meeting the growth of its own load alone has essentially three ways to achieve scale economies: consolidation, wholesale purchases from a larger utility, or coordination. All three have been important sources of rationalization of bulk-power supply throughout the industry's history, and all three have been affected by regulation and other public policies. Before the 1930s, coordination was secondary to consolidation, but the depression and the Public Utility Act of 1935

**Table 3-1. Distribution of Generating Capacity of the 100 Largest
Electric Power Systems, 1964[a]**

| System rank, in descending order | Generating capacity[b] | | Cumulative share of industry subtotal (percent) | Predominantly steam systems |
	Range (megawatts)	Mean		
1–5	13,300–6,995	8,620	21	5
6–10	6,804–4,132	5,263	33	5
11–15	3,872–3,166	3,505	42	5
16–20	3,138–2,849	3,024	49	4
21–25	2,846–2,142	2,650	55	5
26–30	2,058–1,939	2,003	60	5
31–40	1,867–1,446	1,658	68	8
41–50	1,394–1,169	1,262	74	9
51–60	1,148–879	1,002	79	9
61–70	866–687	778	83	8
71–85	669–488	587	87	11
86–100	486–338	397	90	13

Sources and method: Generating capacity data for individual power entities is based on Federal Power Commission, *Statistics of Electric Utilities in the United States, Classes A and B Privately Owned Companies* (1966) and *Statistics of Electric Utilities in the United States, Publicly Owned Systems* (1966). Total industry capacity is from Edison Electric Institute, *Statistical Yearbook of the Electric Utility Industry* (New York: EEI, 1965). The data in the latter are more complete and are consistent with, and largely derived from, FPC data. Assignment of operating companies to systems is based on common ownership of interconnected facilities. Ownership relations are reported in Moody's Investors Service, *Moody's Public Utility Manual, American and Foreign* (1964). Jointly owned capacity is prorated by ownership share.

a. Excludes federal hydro marketing agencies. The 100 systems included represent 90 percent of the total generating capacity in the United States.

b. Generating capacity for the electrical industry in 1964 was 222,285 megawatts, of which 208,057 megawatts were produced by private systems and the Tennessee Valley Authority, and 14,228 by federal hydro marketing agencies. The federal hydro marketing agencies are excluded here because they are best considered as supplemental sources of supply in their respective areas rather than as systems, since they are not responsible for meeting well-defined loads within a service territory, and their capacity additions, as multipurpose projects authorized by Congress, are based on a set of decision criteria somewhat different from the usual power-oriented system planning sequence. TVA is included as a system in the basic industry subtotal since it is the bulk-power supplier for a well-defined service area.

brought an end to the era of holding-company consolidation. Thereafter, coordination became the principal method of rationalization. The breakup of the holding companies, completed in the early 1950s, was followed by a period of remarkable stability in the pattern of ownership, which is reflected in the almost identical size distributions shown in Table 3-2 for 1954, 1959, and 1964. The resurgence of consolidation activity, dating roughly from the formation of Northeast Utilities in 1965, has thus far brought little change from the 1964 size distribution in the industry. Some industry executives, however, predict extreme concentration—perhaps only five to fifteen major systems within the next few decades.

Table 3-2. Distribution of Generating Capacity of Investor-Owned Electric Power Systems, 1954, 1959, and 1964

System rank, in descending order	Mean generating capacity for systems in size group (megawatts)			Cumulative share of investor-owned total (percent)		
	1954	*1959*	*1964*	*1954*	*1959*	*1964*
1–5	3,470	5,259	7,320	21.85	21.28	21.95
6–10	2,233	3,259	4,677	35.91	34.47	35.98
11–15	1,709	2,633	3,358	46.67	45.13	46.05
16–20	1,296	2,230	2,910	54.83	54.15	54.78
21–25	1,104	1,741	2,336	61.78	61.19	61.78
26–30	927	1,394	1,894	67.62	66.83	67.47
31–40	650	1,105	1,491	75.80	75.77	76.41
41–50	489	830	1,159	81.96	82.49	83.36
51–60	385	657	877	86.80	87.81	88.62
61–70	318	469	632	90.80	91.60	92.41
71–85	234	352	431	95.22	95.88	96.26
86–100	133	198	256	97.73	98.28	98.56
101 and below	25	34	60	100.00	100.00	100.00

Source: See note, Table 3-1, for sources and methods of processing. Relevant preceding issues of the sources listed for Table 3-1 were used for 1954 and 1959 data.

The largest ten systems in Table 3-1—the Tennessee Valley Authority and nine private systems—are of particular interest in that they are large enough to absorb units in the scale frontier zone.[12] Interconnections have reduced reserve capacity and captured other network economies for these systems. They have not had to rely heavily on such coordination devices as staggering and joint ownership to increase unit sizes.[13] Among smaller systems the necessary coordination activity varies in-

12. A rule-of-thumb standard of optimum unit size for a large, self-contained system of medium to high density is 7–10 percent of total capacity at the time of installation. The first six to eight systems in the group of ten were large enough to install units in the scale frontier zone in every year from 1954 to 1964 on the basis of a 7 percent rule. Units actually installed by this group during the period were generally close to scale frontiers, the main exceptions being in low-density areas. The basis for the 7–10 percent benchmark is developed in Kirchmayer and others, "Steam-Electric Generating Units," and in Suilin Ling, *Economies of Scale in the Steam-Electric Power Generating Industry* (Amsterdam: North-Holland, 1964).

13. "Staggering" is alternate purchase and sale of power by two or more systems to increase unit size and/or reduce temporary overcapacity. Joint ownership has included a few jointly owned generating subsidiaries and, beginning in the 1960s, direct ownership through tenancy in common.

versely with the size of the system itself and with the size of its neighbors and will be affected by the other coordination options of neighboring systems. Systems in the general range of the eleventh to the twentieth largest might have been able to achieve efficient unit sizes with strong interconnections combined with regular patterns of short-term staggering—bilaterally or in groups of three. A system in the size range of the next forty or so systems would have required either satellite dependency on a nearby large system or minority membership in a tightly organized multilateral pool that planned as if it were a single system.[14] In either case, the system would play a minor role in planning decisions made by the large neighbor or the pool. Systems below the top sixty could achieve efficient performance only as satellites or as weak, dependent members of large pools. Moreover, the smaller the system, the smaller the incentive to existing pools and large systems to accommodate it.

Actual patterns of capacity coordination during the 1950s affected unit sizes primarily through the reserve pooling effect of interconnections and secondarily through limited short-term staggering. In the 1960s there was some use of tenancy in common and increased use of staggering, but a considerable efficiency handicap associated with system size remains.

These inferences, based on observations of coordination patterns, are corroborated by the scale of generating equipment added between 1954 and 1964. The fourteen largest thermal systems provide benchmark ranges for efficient scale under broadly standardized network and load conditions. Statistical analysis[15] suggests that for the smaller systems system size per se was a restraining factor on unit size, although the effect weakened after 1960 as coordination grew. Thus the benchmarks obtained for efficient scale have a downward bias. Even so, the average unit size adopted by medium-density and high-density systems was consistently close to existing scale frontiers.

Comparable statistical analysis of seventy-six major steam-electric systems, other than the big fourteen, showed a very definite association between system peak load and unit size. The relevant elasticity is about 0.4

14. Pools of this order of tightness and scale were rare in the 1950s, but there have been a number of attempts by utilities to develop such pools during the 1960s.

15. Regression and covariance analyses were used. Statistical controls incorporated load-density parameters, peak loads, growth rates, sistership status, network status, annual vintage, and identification of newly pioneered scales. The underlying analytical framework and the research itself are reported in detail in the author's RFF study.

and shows no trend and very low standard errors in all formulations for both annual cross sections and pooled data. In brief, all but a few of the largest systems in the industry have faced a consistent handicap in achieving economies of scale in generation; however, the rapid increase in interconnection and coordination contributed to a rapid increase in scale frontiers and in unit size for systems of every size.[16]

Scale Frontiers since World War II

At the end of World War II, the economic and technical circumstances were ripe for major advances in unit size. An exceptionally large reservoir of technology was built up through the long period of economic depression and war, and economic circumstances also favored large-scale increases in the size of generating units and other power system components. Subsequent progress in advancing scale frontiers has been rapid, but pioneering and early adoption of this technology has been strongly correlated with system size. Regulatory influences on scale-increasing technological change have largely been inadvertent side effects rather than the results of conscious policy, and have had both good and bad effects on industry performance. Regulation might have improved the advance and diffusion of scale-related technology if different merger and coordination policies had been adopted.

The Economic and Regulatory Setting

The principal economic factors that led to a rapid advance in postwar scale frontiers include a war-induced capacity shortage, a rapid load growth, and a cost-price squeeze during periods of inflation.

Load growth reduced the excess capacity resulting from the introduction of larger units and favored interconnection. The effect of both load growth and the war-induced capacity shortage was to necessitate a very large total increase in capacity. Because of the accelerated rate at which new units were being added, experience with new designs and scales could be accumulated quickly. Moreover, the incentive to conserve scarce reserve capacity by strengthening interconnections was especially

16. For the big fourteen and for the seventy-six-system cross section, mean unit size increased at an annual rate slightly greater than that of mean system peak load.

strong. Limited growth during the 1930s slowed advances in scale frontiers, and early postwar scale frontiers represented an unusually small proportion of large-system peaks.[17]

Inflation during World War II, the Korean war, the years from 1956 to 1958, and the late 1960s has created pressures for cost reduction by electric utilities. Because of the regulatory lag, rates are slow to rise during inflation.[18] Before World War II, electric utilities had been comparatively free of such pressures, not only because no general price inflation occurred after the early 1920s, but also because of a favorable rate of technological improvement and the cost effects of load growth. The latter resulted from the addition of a series of off-peak end uses and the consequent improvement in capacity utilization. Though these forces were still at work in the 1950s, past rates of advance in thermal efficiency and capacity utilization could no longer be counted on. Thus, a promising way to reduce generating costs was to focus on major advances in scale.

Regulatory factors also contributed to the pattern of scale increase, probably encouraging advance in scale in some respects and discouraging it in others. Rate regulation was an essential ingredient in the cost-price squeeze, but in moments of candor utility executives have frequently referred to the deterrent effect of rate-base considerations on the willingness of the utilities to depend on wholesale purchases or to engage in long-term staggering. In the early 1960s, when the cost-price squeeze gave way to a period of rising rates of return, rate-base expansion and protection may have assumed increased importance. Part of the attraction of tenancy in common as an alternative to staggering has been that it does not have the rate-base difficulties of staggering. Since the long-run impact of network optimization is decidedly capital-saving, some writers, notably Shepherd, have argued that the Averch-Johnson effect has discouraged interconnection and contributed to excessive reserve capacity.[19] But

17. The 1945 frontier of 75–80 megawatts for unitized designs was only about 4 percent of the average peak load for the five largest steam-electric systems.

18. Under block-rate schedules, average revenue per kilowatt hour declines automatically as consumption per customer increases. This automatic rate decline mitigates and sometimes exceeds the effect of load growth in reducing distribution costs through economies of scale and improved capacity utilization from building off-peak loads.

19. See Harvey Averch and Leland L. Johnson, "Behavior of the Firm under Regulatory Constraint," *American Economic Review*, Vol. 52 (December 1962), pp. 1052–69. See also William G. Shepherd, "Utility Growth and Profits under Regulation," in Shepherd and Thomas G. Gies (eds.), *Utility Regulation: New Di-*

there are potentially offsetting factors. The large, nonrecurring transmission investment involved in major extensions of coordinated networks is attractive to utilities from a near-term rate-base standpoint.

One effect of regulation was the initial avoidance of interstate ties and transmission flows by utilities in Texas, Michigan, Connecticut, and probably several other states to avoid federal jurisdiction. By the mid-1960s, the Texas group was the only serious holdout. On the other hand, federal intervention and the industry's desire to minimize it aided wartime mobilization of the industry, which led to the postwar growth of coordination. Rivalry between public and private power interests can also be shown to have encouraged coordination in some situations while discouraging it in others.

Most relevant, however, is regulation of coordination and mergers, which affects the portion of the industry engaged in frontier-expanding activity. Historically, direct regulation of the terms of coordination agreements, including those affecting rates, has been relatively permissive and has probably had no significant impact on the course of coordination before the 1960s, when both the Federal Power Commission (FPC) and the Antitrust Division of the Justice Department began to take an active interest in access to power pools by systems that had been left out. Informal encouragement of power pooling, which has been a part of the FPC's legislative mandate since 1935, was generally exercised on a modest scale before 1960 and had marginal influence at most. In the 1960s, the FPC actively embarked on a pro-coordination policy—best exemplified by, but not limited to, the National Power Survey. This policy has continued through the present and has probably had a favorable incremental influence on coordination trends that already had considerable momentum.[20]

The lack of major postwar structural improvements through consolidation is probably largely the result of regulation. Postwar merger policy

---

rections in Theory and Policy (Random House, 1966), pp. 12, 51–53. For another analysis of potential Averch-Johnson effects in the electric power industry, as well as a critique of Shepherd's point, see the author's comments in Harry M. Trebing (ed.), Performance under Regulation (Michigan State University, Institute of Public Utilities, 1968), pp. 73–87.

20. For a discussion of the policy rationale involved, see the author's "Regulation and Technological Destiny: The National Power Survey," in American Economic Association, Papers and Proceedings of the Seventy-eighth Annual Meeting, 1965 (American Economic Review, Vol. 56, May 1966), pp. 330–38.

per se posed no explicit barriers, and no important merger attempt during the period was forbidden by regulators. Nevertheless, the industry showed a lack of interest in consolidation after the collapse of the holding company movement and the reforms that followed. The holding company movement itself, especially in the 1920s, had received much of its momentum from the peculiarities, gaps, and internal inconsistencies in the regulatory framework of that time. When the reaction came, it was severe. The industry was understandably sensitive because of its experience during the New Deal and the aura of public suspicion that was part of the heritage of those critical years.[21]

Uncertainty over what the Securities and Exchange Commission's (SEC) jurisdiction over holding companies might mean, and fear of its worst possibilities, certainly contributed to the abandonment of new holding company initiatives until 1965.[22] Even then, when the Northeast Utilities venture was put before a receptive SEC in a comparatively favorable policy climate after the Power Survey, the initiative was considered sufficiently courageous by the industry to merit its prestigious Edison Award. Once the intangible barrier was broken, other consolidation proposals followed quickly, developing into the present movement.

The Technological Setting

Postwar increases in the size of generating units occurred in two overlapping stages. The first step, quickly accomplished within the first generation of units built after the war, was to raise steam conditions (pressure and temperature) and to consolidate prewar and wartime design improvements that favored large units. The second stage was the rapid increase in unit size made feasible and attractive by much higher steam conditions and new unit designs.

Pressure and unit size are closely linked through the substitution of

21. An additional discouraging factor was corporate reorganization and simplification, largely completed by the early 1950s, which usually precluded any new consolidation initiatives for the systems involved while it was in progress.

22. Another factor contributing to the resurgency of consolidation in the 1960s was that by this time many coordination groups had gone as far as they were willing to go with coordination. Power pooling on a "tight" basis has been an important precursor of merger. Northeast Utilities, which sprang from the Connecticut Capacity group and the Connecticut Valley Electric Exchange (CONVEX) operating pool, and the recent merger initiative of two members of the Illinois-Missouri pool are two major cases in point.

capital for fuel in larger units.[23] Power output, other things being equal, is proportional to the kinetic energy of the steam, which in turn depends on the product of pressure and mass. Large increases in pressure can bring commensurate gains in turbine and boiler capacity without enlarging the physical size of the unit very much.

The largest prewar units used low pressures and low temperatures in conjunction with multiple boilers and auxiliaries, frequently sharing these with other units in the same plant. Such complex designs made increasing unit size less attractive in the 1930s than improving steam conditions by introducing reheat cycles or designing better stations by increasing the reliability of boilers and auxiliaries to reduce the number of boilers required for large generators. By 1945, this last development had produced very economical unitized designs featuring one boiler per generator.

The war brought great advances in the metallurgy of high-strength, heat-resistant alloy steels. An 83-megawatt pilot unit installed in 1941 demonstrated the feasibility of doubling pressure.[24] The state of the metallurgical art dictated an emphasis on raising turbine pressure with only moderate increases in temperature. This in turn meant sharply diminishing returns to increased pressure. Doubling pressures from 1,250–1,300 pounds to 2,500–2,600 pounds could reduce heating rates (and fuel requirements) by no more than 10–15 percent; further escalation of pressure would bring much smaller incremental gains in the absence of a breakthrough in handling high temperatures (then unlikely and still unachieved). The new pressure threshold made economical, with very large savings in investment per megawatt, extensions of scale frontiers. By the mid-1960s, when the frontier had passed 500 megawatts, 2,400 pounds per square inch was still the prevailing pressure for the largest units, even though 3,500 pounds per square inch was available.

The complementary technologies of transmission and load-frequency control permitted, and perhaps even encouraged, an accelerated movement to larger units. Progress in transmission technology has been a consistent, long-term evolution—timed in relation to load growth and unit sizes. This contrasts with the false image of dramatic advances often con-

23. Thermal efficiency is largely a function of temperature and pressure, although there are minor, pure size effects. For a discussion of the interactions of capital, fuel, and scale, see Ling, *Economies of Scale*, pp. 28–32.

24. This unit at the Twin Branch plant of the American Gas and Electric Company (now American Electric Power Company) provided experience on pressures of 2,000–2,500 pounds per square inch.

veyed by popular news reports on recent extra-high-voltage (EHV) developments, long-distance interties, mine-mouth plants, and direct current technology. Latent scale economies condition developments in transmission even more than in generation. The dominance of the latter in bulk-power costs has a strong governing effect on scale frontiers in transmission.

From the 1920s through the 1950s, a series of design improvements in established voltage levels (220 and 110 kilovolts) brought substantial cumulative improvements in performance in terms of increased capacity, reliability, and feasible distances. These levels proved adequate for economically incorporating frontier-sized units into stable, reliable networks during the 1950s. The timing of pioneering work on EHV developments (345 kilovolts and up) has closely reflected the anticipated demand for high-capacity and long-distance circuits. Transmission technology was not a critical bottleneck in increasing the scale of the generating units.

THERMAL FRONTIERS. Thermal frontiers (that is, technically feasible steam conditions) were pushed much more aggressively after the war than before. Plans for the pilot unit for the second major postwar escalation of steam conditions—supercritical pressures—were announced in 1953,[25] after only a few years of commercial experience with the newly pioneered level of 1,800–2,000 pounds per square inch. Moreover, construction of the pilot unit had scarcely begun when a second utility announced plans for two large supercritical units involving a major extrapolation of scale as well as unusually high pressures and temperatures.[26] Several more supercritical units were committed before these units were in operation. By the early 1960s, the commercial feasibility of 3,500-pounds-per-square-inch units had been established, and more important, the economic limits to postwar advances in steam conditions had been probed within a very short period.

The high rate of introduction of supercritical units was achieved at the

25. This unit, installed at the Philo plant of American Gas and Electric, was conservatively scaled at the 100- to 125-megawatt range, well within the relevant frontier of over 200 megawatts and far below the size at which supercritical pressures would be optimal. Steam conditions featured a pressure of 4,500 pounds per square inch and an initial temperature of 1,150 degrees. At supercritical pressures water converts to steam without boiling, which saves energy.

26. These units, installed at Philadelphia Electric's Eddystone plant, were 325 megawatts each. Pressures were 5,000 and 3,500 pounds per square inch, respectively.

price of increased maintenance troubles and reduced service time.[27] Some of the problems experienced with early vintage supercritical units "might have been avoided if more experience had been available from Philo."[28] Nevertheless, the consensus on the overall performance of this generation of units has been favorable.

As of the mid-1960s, there was no general trend toward adoption of supercritical pressures comparable to the rapid movement to pressures in the 1,800- to 2,400-pounds-per-square-inch range. The reasons are economic. In most cases, expected gains in thermal efficiency were not large enough to justify the extra investment.

The best practice in the 1960s featured a range of pressures above 2,000 pounds per square inch for large units, how much above depending on individual economic circumstances. Steam conditions and design refinements had reached levels at which theoretical gains from further development were comparatively small without large temperature increases. With nuclear power on the threshold of widespread adoption and the expansion of scale frontiers in full swing, pioneering after 1960 focused on these more promising areas.

Table 3-3 presents the statistical history of pressure frontiers as they expanded upward from 1,250 pounds per square inch and data on the participation of large utility systems. The first five to ten units of each major advance in pressure required only a few years. The number of utilities actively involved during the pioneering of the main advances in pressure is small, and the very large systems are disproportionately represented, principally because the economics of unit size and pressure calls for a close association between the two.[29]

SCALE FRONTIERS. In view of the state of technical knowledge in 1945

27. Eddystone Number One is an extreme example. It was out of service for most of 1963, the year after the plant had set a new thermal efficiency record.

28. *National Power Survey*, Pt. 2, p. 52. Copper deposits in the turbines are mentioned as a common problem.

29. Nonscale factors such as fuel prices and managerial judgment about technical risks also help determine the most economical pressure. Consequently, large systems are less dominant in thermal innovations than in extending scale frontiers per se. Nevertheless, the key frontier levels of 2,000, 2,400, and 3,500 pounds per square inch show that large systems predominated during the critical first few years of introduction. Nearly all of the systems, other than the big fourteen, that participated in the introduction of new scales and pressures were among the twenty to twenty-five largest.

Table 3-3. Thermal Frontiers in Pressure Levels for Electric Power Systems, 1935–65

Pressure class[a] (pounds per square inch)	Period of introduction	Units added	Units added by 14 largest systems	Systems involved	Number of systems among 14 largest
1,250	1935–38	5	3	5	3
	1935–45	23	21	8	5
	1935–48	39	29	13	8
	1935–52	116	48	35	10
1,450	1948–49	3	3	1	1
	1948–50	8	6	4	2
	1948–52	42	20	22	7
	1948–54	96 (100)[b]	34	36[b]	9
1,800	1949	1	0	1	0
	1949–52	6	3	3	1
	1949–54	50	38	17	11
	1949–56	95	66	30	12
2,000	(1941)[c]	(1)	(1)	(1)	(1)
	1949–50	3	3	1	1
	1949–54	11	11	1	1
	1949–56	25 (31)[b]	13	7[b]	2
	1949–58	50 (56)[b]	16	10[b]	2
2,400	(1941)[c]	(1)	(1)	(1)	(1)
	1953–55	4	3	3	2
	1953–58	19	10	12	7
	1953–60	47	20	20	8
	1953–62	73	30	25	9
3,500	1957	1	1	1	1
	1957–60	5	4	3	2
	1957–62	6	5	3	2
	1957–65	10	7	5	3

Source: Federal Power Commission, *Steam-Electric Plant Construction Cost and Annual Production Expenses*, relevant years.

a. All pressures are for the initial turbine stages. The nominal pressure for each pressure class is that adopted for a majority of the units in the class. Each class represents a range of pressures (in pounds per square inch—psi) as follows

1,250 from 1,200 to 1,350
1,450 from 1,450 to 1,500
1,800 from 1,650 to 1,900
2,000 from 1,950 to 2,150
2,400 from 2,200 to 2,600
3,500 for all supercritical units

Topping units were excluded from the tabulations.

b. Four 1,450 psi units of Electric Energy, Inc., and six 2,000 psi units of the Ohio Valley Electric Corporation are included in parentheses. Each of the two firms is owned jointly by a group of utilities and is not included in the regular tabulations of units or systems in the table. None of the owning systems of EE is among the big fourteen, but the largest stockholder (40 percent) in OVEC was American Gas and Electric, which dominated the pioneering of the 2,000 psi level.

c. The pilot unit at the Twin Branch plant of AG&E is shown in parentheses for both the 2,000 and 2,400 psi classes.

and of the low ratio of the maximum unit size (75 megawatts) to large system peaks at the end of the war, it is not surprising that scale frontiers at several levels, ranging from 100 to 150 megawatts, expanded almost simultaneously after the war. The data in Table 3-4 testify to the rapid and continuing increases in unit size that followed.

By the mid-1950s, the war-induced capacity shortage had been largely made up, and with the scale frontier growing faster than peak loads, the number of units installed annually declined. Whereas a total of sixty-two units of 100 to 175 megawatts were installed from 1950 through 1953, only thirty-four units of 189 megawatts and above were installed during the comparable first four years after their introduction, 1956 through 1959. As a result, further large increases in unit size proceeded on the basis of progressively more limited experience with immediately preceding scale frontiers.

To maintain the pace of advance, the industry had to eliminate, or else design around, an increasing number of bottlenecks. Where bottlenecks or technical limits could not be eliminated immediately, designs became more complex, with some loss of scale economies. An obvious example is cross-compounding.[30] While maximum unit size was below 225 megawatts, there was virtually no gap between the largest cross-compound and the largest tandem-compound turbine-generator sets purchased, but as the frontier moved to larger scales, the gap widened to a ratio of more than two to one by the early 1960s.[31]

Removal and circumvention of technical limits and bottlenecks is en-

30. Other examples include (1) duplication of boiler tubing and other small components to achieve large flows, (2) adoption of nonoptimum dimensions for large components to avoid exceeding transport clearances, and (3) expensive field assembly of components too large to be transported.

31. The 1,000-megawatt unit installed at Consolidated Edison's Ravenswood plant in 1965 uses two boilers, as well as cross-compounding, to achieve its capacity. In the mid-1960s, increasing the size of tandem-compound units became an attractive alternative to enlarging cross-compound units. For example, the American Electric Power Company, which played a major role in expanding earlier postwar cross-compound frontiers at 125, 175, 225, 450, and 580 megawatts, moderated its emphasis on increasing overall unit size in the early 1960s in order to develop improved designs in the 600-megawatt range. While the other large systems were pushing cross-compound frontiers into the 900- to 1,100-megawatt range, AEP chose to extend the frontier for tandem-compound units above 600 megawatts. Transmission shows a similar alternation between simplification and complication as scale frontiers increase. For instance, banks of single-phase transformers are substituted for three-phase transformers when the size limits of the latter are reached; and circuit breakers of feasible design are cascaded under similar circumstances.

Table 3-4. Scale Frontiers in Steam Units for Electric Power Systems, 1935–65

Scale class[a] (megawatts)	Period of introduction	Units added[b]	Units added by 14 largest systems	Systems involved	Number of systems among 14 largest
80	1935	1	0	1	0
	1935–41	4	3	3	2
	1935–45	7	4	4	2
	1935–49	15	—	—	—
100	1947–50	8	8	4	4
	1947–51	14	13	6	5
	1947–52	21	18	9	6
125	1949–51	10	10	4	4
	1949–52	18	18	4	4
	1949–53	28	27	8	7
150	1948–53	6	6	2	2
	1948–54	14	12	6	5
	1948–55	21	15	11	7
165	1950	3	3	1	1
175	1953–55	33	33	5	5
	1953–56	36	35	6	5
	1953–57	37	36	6	5
200	1955–57	9	8	6	5
	1955–58	13	12	7	6
	1955–59	19	17	10	7
250	1956–58	7	4	5	3
	1956–59	20	10	11	6
	1956–60	24	13	14	8
325	1958–60	12	12	6	6
	1958–61	21	19	10	8
	1958–62	23	20	11	8
400	1960–62	9	8	6	5
	1960–63	14	13	9	6
	1960–64	21	17	10	6
500	1961–63	4	4	2	2
	1961–64	5	5	3	3
	1961–65	9	9	5	5

Footnotes for this table are on the next page

demic to all extensions of the scale frontier, and the historical record is full of examples of apparently intractable limitations that later turned out to be temporary. Nevertheless, increased scale is inherently associated with increased complexity of design and with the use of compromises to get around the more difficult size limitations. The "consensus" cost function, with rising cost per unit of output as scale increases beyond 400 megawatts, reflects this tendency.

Since 1965, scale frontiers have continued to expand, although the pace is apparently more modest than in the 1950s. The pattern is complicated by the transition to nuclear units. Orders have been placed for 1,300-megawatt units, the 800- to 1,100-megawatt range has lost its novelty, and technologists continue to show ingenuity in reducing scale-related design problems to tractable dimensions.

The close association between system size and choice of unit size accounts for the dominance of large systems in scale pioneering. The data in Table 3-4 indicate that nearly all of the first ten to fifteen units in each scale class were added by the fourteen largest systems.

Appraisal of the Record

The postwar record suggests two principal conclusions about the impact of industry organization on relevant aspects of performance. First, the main potential for improvement appears to be in the rate of adoption of recently pioneered pressures rather than in the acceleration of major

Source: Federal Power Commission, *Steam-Electric Plant Construction Cost and Annual Production Expenses,* relevant years.

a. Figures for scale are nameplate ratings as reported in ibid. Scale classes embrace the respective ranges as shown below. In all cases, the majority of units are clustered at or near the nominal scale.

Scale class (megawatts)	Range (megawatts)
80	75–85
100	100–110
125	112.5–135
150	145–156
165	165
175	169–180
200	200–225
250	231.3–275
325	290–335
400	359–450
500	526–1,028

The last figures in the second column include one at 1,028, two at 704, one at 660, and five in the 500–600 range.

b. Only units with single boilers are included in the tabulations. Experience through 1947 with large multiboiler installations includes eighteen units of 75–80 megawatts, starting in 1928; ten units of 100–115 megawatts, starting in 1928; ten units of 125–165 megawatts, starting in 1929; and one 208-megawatt installation in 1929.

new technological developments. Second, both the locus of pioneering and
the rate of subsequent adoption of newly available frontiers were clearly
influenced by the size distribution of electric power systems and by imper-
fections in coordination.

The first five to ten units of each major postwar advance in pressure
required only a few years, and the move to supercritical pressures was
made so aggressively that they were probably available commercially in
advance of any widespread demand that might have developed in a fully
coordinated, or rationalized, industry. In view of this high absolute rate
of advance, the importance of time-related aspects of experience, and the
low thermal efficiency payoff from pressures above 3,500 pounds without
a major metallurgical breakthrough, it appears that industry performance
in advancing the thermal frontier (given the situation before 1945) was
close to the potential under the most favorable combination of regulation
and industry organization.[32]

Progress in expanding the scale frontier was not as swift. Expansion
might have proceeded more rapidly if the size mix of new units had been
determined by the demands of a rationalized industry. The technological
progress needed for expanding unit size was the removal of numerous
design bottlenecks, which individually were minor and which probably
could have been overcome if design and construction experience had
been accumulated at a faster pace. The procurement practices of large
systems and the nature of new capacity being added indicate that the annual demand for the frontier-range units would have at least doubled if
there had been no institutional restrictions on choice of scale by individual utilities. How much difference this would have made in advancing
scale frontiers is a matter for speculation, but the direction of the effect is
clear.

The diffusion of newly established scales and steam conditions would
clearly have been more rapid in a rationalized industry. This is evident
from the dominant role that the large systems played in pioneering and
adopting frontier scales and designs, as well as from the observed corre-

32. The U.S. record also compares very favorably with that of the large foreign systems. See F. P. R. Brechling, "Techniques of Production in the British,
French, and U.S. Coal-Fired Electricity Generating Industries: A Comparative
Study" (London: National Institute of Economic and Social Research, 1965;
mimeographed). See also Philip Sporn, "Soviet Shift to Big HP Steam Units Holds
No Surprises," *Electrical World* (May 29, 1961), pp. 50–55, for comparative technical data on the Soviet and U.S. industries.

lations among system peaks, unit sizes, and pressures for choices within established frontiers.

Policy Considerations

The aspects of economic performance discussed in this chapter are clearly related to the organization of the industry. This can be influenced to some extent by regulatory policy, especially that concerning mergers, coordination, and wholesale rates.

The essence of the practical policy problem (which excludes large, utopian reforms) is how to influence the evolution of the industry to improve scale-related performance without unduly compromising other policy considerations. The present movement toward consolidation and coordination is an evolutionary trend with considerable momentum. Public policy to date has, on balance, encouraged this trend—especially the active support by the FPC of coordination and the accommodative, though noncommittal, position on mergers of the FPC and SEC. Nevertheless, policy in both of these areas is still somewhat plastic, making timely some fundamental questions about the relations between industry structure and the goals of policy.

How do past structure and performance patterns relate to the future? How much improvement in performance can be expected from current industry trends within the present policy environment? What are the possibilities for accelerating the growth of coordination and merger to hasten achievement of a better mix of coordination and integration? What, if any, sacrifices in scale-related performance would be made if consolidation were restrained by a policy that sought to avoid concentrated ownership? How large must the basic planning units (systems or "tight" pools) be to optimize network performance? The following remarks touch these questions only briefly, suggesting some relevant considerations rather than trying to offer definitive answers.

Structure and Performance to Date

The evidence indicates that industry performance has been closely related to the size distribution of electric utility systems. The choice of unit sizes has been closely associated with system size. A few very large systems account for most sales of scale-frontier-size units and thereby play a dominant role in scale pioneering. A rationally organized industry would certainly have more systems large enough to demand scale-frontier units. This would probably push out scale frontiers further than they have reached today, although the magnitude of this effect is open to speculation. The efficiency loss caused by small systems purchasing smaller units (the scale gap) and from a suboptimal degree of reserve capacity pooling is perhaps 4 to 10 percent of total bulk-power supply costs.

The scale gap was comparatively stable from the early 1950s through the mid-1960s despite impressive growth in interconnections and power pooling. Coordination activity by large systems helped the growth of scale frontiers; coordination by smaller systems merely prevented their relative position from deteriorating. In the 1960s the growth of power pooling has probably accelerated, though perhaps not as much as increased publicity and governmental attention might indicate. Just as coordination tended to be underrated in the years before the Power Survey when there was little publicity, the more dramatic and impressive aspects of current coordination now tend to be incorrectly viewed as typical of general industry performance.

Important evidence pertaining to the various "superpools" that have been formed during the 1960s is only now accumulating. In the 1950s and early 1960s, only a few multilateral power pools approached fully integrated capacity expansion. Each of these pools consists of only three or four systems and has a total capacity smaller than the largest systems in the industry.[33] The superpools involve much larger blocks of capacity and many more systems. Their early announcements about capacity, while providing insufficient information for judging their effectiveness in integrating system planning and capacity additions, suggest that improvements have been made during the 1960s. For instance, more area-wide EHV backbone grids have been planned, and jointly owned generating units are much more common than in the past. Despite the probable effect of recent actions, a broad association between unit and system size continues to be evident from data on capacity additions and plans, and the

33. The largest multilateral capacity group that had reached this degree of "tightness" by the early 1960s was the Illinois-Missouri Pool, a three-system group (prior to the merger of two members) with a combined capacity approximating that of the tenth largest system in the industry. The pool adopted unit sizes similar to those of systems of that general size. The other comparably tight multilateral groups were much smaller; no more than five such groups (and probably fewer) were among the hundred largest systems.

implicit scale gap remains substantial. A closer assessment of the amount of actual improvement since the mid-1960s awaits further research.

Future Relevance of Scale-Related Performance

If the future of the electric power industry were tied to conventional fossil-fuel technology, the gap between actual and potential generating cost might be expected to diminish gradually. Even with the size distribution of utility systems and the degree of coordination both unchanged, a constant mix of unit sizes would imply a declining percentage gap in costs, since the growth of existing utility systems would move them progressively into scale ranges where cost elasticities are 0.9 and above. Furthermore, both coordination and the current merger movement already have enough momentum to achieve some gap closing, even if the regulatory climate remains unchanged.

Several developments, however, suggest that the organizational problems connected with scale-related performance will continue to be important. These include the transition to nuclear power, environmental considerations, and the continued importance of scale-related technological progress in transmission. The question of the reliability of large networks is complementary to the other aspects of the coordination problem.

NUCLEAR DEVELOPMENTS. Nuclear power technology may increase the potential scale-gap problem for generating units. Most experts believe that economies of scale in nuclear power are greater than in conventional steam power because of very marked scale economies in the reactor and heat-exchange stages. (The turbines and generators have much the same scale economies in both nuclear and conventional units.) The elasticity of costs with respect to unit size for the few engineering functions thus far observed for nuclear units is about 0.7 for all unit sizes up to more than 3,000 megawatts. The largest sizes are considered feasible within the next two or three decades. Breeder technology favors larger units than reactor designs currently used.

It would be unwise to accept this prognosis uncritically. While each reactor design has substantial built-in economies of scale, scale frontiers vary according to the type of reactor, and the comparative newness of nuclear technology affords opportunities for reducing costs by improving design for a given scale.[34] Shifts in cost functions for a given unit size

34. In a comment on the preconference draft of this chapter, Paul MacAvoy expressed the view that focusing on reducing costs for reactor design with rela-

reduce the relative attractiveness of pushing out the scale frontier. The focus of research, development, and early experience with nuclear technology may have a great influence on the extent to which the electric power industry is faced with a scale-gap problem. The size of the gap will also depend on the actions and policies of the Atomic Energy Commission, the manufacturers' perception of technological opportunities, and the choices of the electric utilities themselves. These choices will most certainly reflect the organization of the industry, and it is important that nuclear options be examined from this standpoint.

TRANSMISSION DEVELOPMENTS. Scale economies in transmission can be expected to remain important. These economies, as well as expected progress in both extra-high-voltage and ultra-high-voltage technologies, should tend to favor large planning units and perhaps reduce the transmission penalties caused by larger units.

ENVIRONMENTAL CONSIDERATIONS. The optimal scale mix of future generating and transmission system components will probably depend to a large extent on environmental considerations. These include conservation of scarce land and water resources for plant sites and rights-of-way. Environmental considerations create organizational requirements in planning power systems that essentially parallel, reinforce, and possibly even go beyond those for achieving efficient unit size. In addition, the scales of generating units and transmission components are probably increased, on balance, when environmental factors are given weight in coordinated decision making.

A socially efficient trade-off between environmental values and power supply costs calls for a more thoroughgoing coordination effort than merely approximating minimum commercial bulk-power costs. The latter can be approached by patching together locally planned transmission networks with heavy interconnections and backbone grids and by arranging for large units without considering very carefully their geographical distribution. The result is redundant use of rights-of-way and plant sites, as well as inferior locations and routing.[35] Changing the network to re-

---

tively intractable scale frontiers might well be a more promising approach than extending frontiers to realize latent scale economies. He is also skeptical about engineering estimates of scale economies for application far in the future and suggests the presence of an optimistic bias in predictions of this kind.

35. In power pooling, the geographical distribution of capacity is often, though by no means always, influenced by a desire to prevent large deficits in capacity in the area covered by each principal member of the pool.

duce unfavorable environmental side effects at minimum cost requires close evaluation of the full range of siting, routing, design, and other alternatives for geographical areas larger than those covered by the service territories of the largest medium-density and high-density systems.

The need for optimal use of scarce urban and waterfront plant sites, as well as the scarcity of land for right-of-way and the increased emphasis on selective undergrounding, means that new capacity will have to be concentrated in fewer units and stations and high-capacity transmission circuits will be needed. Though air and thermal pollution, safety, and other environmental considerations will undoubtedly constrain station size at some sites, the effect can be either to increase or to decrease desired unit size. The need to locate more capacity at alternative sites puts a premium on size at those locations.

Alternative Policy Choices

Network optimization might be approached through a variety of adjustment patterns, each relying on a different combination of increased coordination, increased wholesale-retail specialization, and consolidation. No adequate foundation has been laid in this chapter for a constructive discussion of actual policy choices, but the structure-performance analysis developed here is relevant to these choices.

For any pattern of ownership and wholesale-retail division that is likely within the next two decades, greater coordination than in the past will be needed to approximate network optimization, especially if adjustments for environmental factors are effectively to take into account the range of options available to the network. Consequently, consolidation and coordination may be complementary forces for rationalization as much as they are competing alternatives, especially since merger frequently grows out of coordination. It is helpful to envision merger policy as constraining and guiding the long-term development of industry organization, while at any given time the ownership pattern determines the amount of coordination effort or wholesaling, or both, then necessary for network optimization.

The policy approach that I favor, asserted only loosely here, calls for encouraging adjustment on all three fronts within broad organizational constraints designed to assure the consistency of the ultimate industry structure with the goals of policy. The means of choosing a target, or end-goal, industry structure within the investor-owned segment, and some policy considerations relevant to it, are discussed below.

An Organizational Benchmark

A comparison of possible end-goal industry organizations from an overall performance standpoint is facilitated by adopting an analytical benchmark, or "base case." Ideally, such a benchmark should have several characteristics. First, it should be expressed in simply defined organizational units that have well-understood economic performance characteristics. Second, without losing objectivity and simplicity, it should allow for the variety of concentration-coordination mixes that might be consistent with efficient and progressive network expansion. Finally, a well-designed benchmark should facilitate intelligent treatment of trade-offs between bulk-power rationalization and other policy considerations.

The variety of possible target organizations stems from two sources. First, the minimum optimal system size for any feasible degree of coordination might be exceeded by some or all systems without any loss in scale-related performance. Second, coordination at its best might achieve the same network integration in some areas and over some ranges of system size as would consolidation. To attain large-system performance, a group of smaller systems must make basic planning and operating decisions on a combined, one-system basis. An organizational benchmark can therefore be defined in terms of "planning units" (which may be either systems or high-performance, tight pools) without prejudging the number of cases in which coordination can achieve efficient results.

The organizational benchmark adopted here is the minimum optimum scale associated with a level of coordination already shown to be feasible.[36] The increased cost of power associated with smaller planning units, in combination with the intra-unit coordination that has been achieved, can be measured.[37]

---

36. Minimum optimum scale should be interpreted dynamically as growing with demand and the extension of scale frontiers. The optimum embraces both the adoption of efficiently large system components and the advancement of the scale frontier at the most economical rate.

37. Measurements of this sort are a straightforward byproduct of the author's scale-gap research, but publishable estimates are not yet available.

Broad policy considerations affecting the organization of the power industry are numerous and are often closely linked with a panoply of complicated issues in overall power policy.[38] The history of policy debate is rich with confusion (sometimes desired by the participants) over the relation of broad considerations to more narrowly economic aspects of performance. Unless broad policy considerations are defined in advance and their organizational implications identified, there is likely to be a tendency to adopt whichever of two evil extremes best fits one's predisposition. The efficiency-minded may disregard broader considerations as nebulous and lacking a demonstrated relation to industry organization, while others may introduce a full array of issues, without explicitly analyzing them, as grounds for denying the importance of benchmarks based on economic performance.

The broad policy concerns that might have organizational implications fall into two large groups: "neutral" and anticoncentration. Concentration beyond levels that promote efficiency and progress has little to recommend it; therefore efficiency considerations should set a rough upper limit to desirable concentration. Whether lower levels of concentration might be preferred depends on the trade-off between higher power costs and expected benefits from less concentration.

The approach taken here is to define the benchmark and discuss some organization-related policy considerations that enter into the relevant trade-off without proposing a solution or covering the full range of relevant issues.

38. One of the most highly charged and hard-to-treat areas envelops a wide range of interrelated public-private power issues—competition for industrial customers, territorial disputes, legal differences in tax and financing status, and preference granted to publicly owned and cooperative systems in buying from government hydro projects. This is only a partial list of examples. These public-private issues interact with and overlap various broader regulatory questions, including licensing of power facilities, wholesale rate regulation, antitrust aspects of power pooling and wholesaling, issues relating to the use of one utility's transmission lines by another, and access to power pools by uninvited systems—the latter usually being very small and often publicly owned. These issues, and others not listed, are tightly linked in terms of their narrowly economic effects; moreover, they tend to become confused and intermixed in the highly charged politics of electric power. For instance, much of the debate over the various "reliability" bills is concerned with such questions as those listed above. These matters, especially important from the standpoint of economic justice, have much less impact on overall industry efficiency and technical progress, which are related to performance within the (largely private) major system group.

An industry with twenty to thirty major planning units[39] of approximately the same size appears to meet our benchmark standard. It has a size range of planning units equivalent to that of the five or six largest U.S. systems.[40] All of these systems have been able to absorb existing scale-frontier unit sizes comfortably (wherever they are economically justified) on the strength of their own loads and the partial reserve pooling effect of their interconnection. Their prospective performance as the optimal scale frontiers become higher, particularly if nuclear scale economies prove to be very great, is more uncertain. Well-established (though not universally adopted) coordination practices, however, could have doubled the scales the largest systems have adopted if such scales had been available and economically attractive. Strong interconnections and integrated planning of facilities could permit unit sizes of around 20 percent of individual planning unit peaks, a very large size in relation to what is technically feasible and economically justified.[41] The essential point is the feasibility of coordination to obtain such scales, *if they are economically justified,* in a benchmark industry. In view of time-related limits to scale frontiers, a benchmark industry would probably be concentrated enough to develop and adopt any rate of advance in scale frontiers that might reasonably be expected to be optimal.[42]

Even if the existing rate of advance in scale frontiers proves desirable economically, and even if consolidation proceeds to the point of only one investor-owned utility per planning unit, an optimized benchmark industry would involve very substantial coordination effort, both for achieving network economies at interplanning unit levels and for coordinating the predominant investor-owned utilities with public and cooperative systems.[43]

39. The major planning units would not necessarily include the entire industry. Smaller systems would serve remote and low-density areas.

40. A twenty-unit industry implies an average size of major planning units slightly smaller than the Tennessee Valley Authority (TVA), the largest U.S. system. A thirty-unit industry implies an average-sized planning unit slightly smaller than the sixth largest U.S. system.

41. Early 1960s scale frontiers ranged around 400 to 600 megawatts—less than 5 percent of TVA peaks at the time and less than 10 percent of the peaks of the fifth and sixth largest systems.

42. In the light of new evidence on future scale frontiers and feasible coordination, this benchmark might be adjusted upward or downward.

43. A reasonable assumption in any realistic policy is that the various ownership segments are likely to coexist for the indefinite future, with their relative shares changing slowly and incrementally, if at all. Although "open pools" face

Considerations for Merger Policy

If the regulatory authorities are assumed to be going ahead rapidly with appropriate coordination and wholesale-rate policies, how might merger policy be used to further development of an end-goal industry structure?

To date, merger policy at all levels (in the SEC for holding companies, the FPC for interstate utilities, the Antitrust Division of the Department of Justice as a guardian of competition, and some state commissions, within their jurisdictions) has been a mixture of accommodation and selective resistance. Initiative lies with those utilities wanting to merge. In some state jurisdictions, mergers are permitted unless an adverse effect on the public interest is shown. The FPC and the SEC require an affirmative showing of consistency with the public interest. In general, neither the law nor the context in which policy decisions are made has clearly encouraged regulators to consider each merger proposal in relation to alternative merger possibilities from the standpoint of their impact on overall industry organization and performance. Thus far in the current wave of consolidations, no important merger application has been blocked, but signs of increasing public policy resistance are appearing. Conditions (for example, spin-off of gas properties) have sometimes been attached to approval, and it has been suggested that for antitrust reasons, approval of mergers by very large systems may be increasingly difficult to obtain in the future. More important, the Antitrust Division has intervened in at least two merger cases, and the SEC staff has opposed the acquisition of the Columbus and Southern Ohio Electric Company by American Electric Power.[44]

---

very difficult problems of effective implementation, and public policy in this difficult area is in a confusing state of contention, some sort of working arrangement featuring a mixture of open pooling and wholesaling seems to be evolving. While this issue carries heavy weight in current regulatory and antitrust action and involves complex and important problems of economic justice for those affected, it should be noted that the share of generating capacity currently accounted for by public and cooperative systems is extremely small.

44. For surveys of these developments, see "Feds Taking Tougher Line on Utility Consolidations," *Electrical World* (Feb. 16, 1970), pp. 21–22; and "Justice Department Takes Hard Line on Mergers," *Electrical World* (May 18, 1970), pp. 19–20. For the most part, past regulation of mergers has been concerned with their financial and accounting aspects, particularly the welfare of affected security holders and the prevention of write-ups that would increase the rate base. Con-

Network integration is mentioned frequently in both the Federal Power Act and the Public Utility Holding Company Act, but only in the latter does it appear explicitly in the merger section.[45]

The current movement has involved mergers by contiguous systems, well located for consolidation with one another. In only one or two minor cases were alternatives clearly superior from a system integration standpoint. Mergers thus far have been largely consistent with the creation of a more symmetrical size distribution of the industry, but the acquiring systems in recent or proposed major acquisitions include such large utilities as American Electric Power, Commonwealth Edison, and Southern California Edison. None of these creates a system with a peak load as high as 5 percent of the industry total. The twenty- to thirty-system benchmark used here suggests that the five or six largest systems in the country, including the first two mentioned above, are likely to be large enough for incorporation into a rationalized industry structure without further merger.

Beyond progressiveness and efficiency, several other factors should be considered in making merger policy. First, flexibility should be preserved. Actions that constrain future decisions should not be taken if no recognized positive gain results. Second, proximate decisions should be compatible with ultimate goals. A consolidation that directly improves the performance of the merging firms may still be inferior to alternative patterns of amalgamation. Third, the effects on competition should be considered. These are hard to evaluate in the regulated sector, but they are present and must be weighed. Fourth, industry initiatives and historical relationships, as reflected in existing interconnection patterns and power-pool affiliations, are sometimes useful and should not be ignored.

---

cern that the acquiring company might be "too big" is occasionally voiced in the statements and opinions of regulators, and the basis for these statements seems to lie in antitrust ideology dealing with large size. However, these expressions of concern have been just that; the specifics of individual decisions (except possibly the disposition of gas properties) have been the same as if the comments about bigness had never been made. One has to go back to the breakup of the holding companies in the 1930s and 1940s to find a completed case in which a proposed merger was blocked because the acquiring company was believed to be too large for further acquisitions at the time. The one important case of this type that arose during the breakup period involved the same two utilities as did the proposed affiliation recently challenged by the SEC staff.

45. The Public Utility Holding Company Act goes even further in its Section 30, which has been largely unused. The section provides for SEC studies to determine what ownership pattern in each area would serve the public interest most efficiently.

Fifth, interfirm comparisons in the regulation and management of electric utilities can play an important role in judging efficiency. Sixth, since investor-owned utilities are usually vertically integrated, the impact of bulk-power mergers on the organization of distribution should be considered.

FLEXIBILITY AND COMPATIBILITY WITH END GOALS. Flexibility and compatibility with ultimate objectives are closely related. They imply that mergers should not be undertaken if no clear gain in performance is expected and if local efficiency gains for the utilities immediately involved do not ensure that a merger will improve the performance of the industry as a whole. Reorganizing an industry involves a great deal of time and trouble, and the efficient and progressive industry structure is one that is well adapted to a long-term balance among load growth, optimum scale of equipment, and extension of scale frontiers. Consolidations usually (and most conveniently) take place among lumpy units of entire power systems. These indivisibilities determine the feasible levels of concentration, and concentration in excess of efficiency requirements could limit the number of efficient systems if further concentration is required by future technical change. For example, breeder reactors may have major economies of scale; these might justify eventual increases in power industry concentration beyond the level that now seems appropriate. Patterns of asymmetry in size can also limit sharply the ultimate optimum number of major systems. There are some examples of this in the history of railroad consolidation. If the current merger wave were to result in, say, eighteen large systems of uneven size plus a geographically scattered residual of much smaller utilities, it might be necessary to go to as few as eight or ten systems to attain the minimum sizes achievable in a symmetrical size distribution of about twenty systems.

A related point is that it is much easier and less costly to prevent a merger than to undo it once it has taken place. The reasons for preferring the unmerged state may be more apparent later than they are beforehand. Pieces of property that are bypassed during a merger movement's early thrust eventually get tacked onto systems in their own vicinity. The result of this sequence of merger is far fewer systems than the maximum that is consistent with an efficient bulk-power organization.

COMPETITIVE CONSIDERATIONS. Competition in the power industry appears mainly as rivalry within the privately owned sector in purchasing inputs and in attracting industrial customers, rivalry between privately and publicly owned systems, and rivalry among competing energy forms. Nonmarket rivalry also arises, and is discussed here in connection with the role of comparison in regulated industries. In general, the larger the number of major power systems, the more competition there is, but the types of competition are not homogeneous in their normative effects.

The issues relating to competition in the purchase of inputs (fuel, equipment, construction, and engineering services) are basically like those posed for unregulated industries.[46] National and international markets prevail in equipment, construction, and engineering, with generally high seller concentration; if the buyers consisted of twenty to thirty large systems the present competitiveness of these markets would certainly not be reduced and might even be enhanced.[47] Fossil-fuel markets, especially the market for coal, tend to be regional, giving very large systems substantial market power. Because of economies of scale in fuel transportation, only a few buyers and sellers compete; however, fuel and fuel transport markets may decline in importance as the nuclear share of the industry increases.

In the case of other types of competition the implications for merger policy are cloudier. The normative consequences of competition among franchised monopolies, which may frequently find it profitable to sell below marginal cost in competitive submarkets, are considerably less clear than in the case of unregulated firms. Perhaps the beliefs that really underlie a strong pro-competitive stance have at least as much to do with the image of competition as an invigorating, jarring antidote to the quiet life as with the adjustment to Pareto optimal equilibrium. Whereas the latter aspect of the competitive norm does not necessarily favor competition among franchised monopolies, the former aspect does.

Finally, consolidation in the electric utility industry may affect public-private power rivalry and competition between electricity and gas. The total effect of mergers on these forms of competition is not immediately clear, but the issues involved should be closely analyzed in formulating merger policy.

46. Westfield's "joint conspiracy" thesis suggests that a given level of concentration may mean less competition in the purchase of capital inputs than in the case of unregulated industries. This would make the case against concentration stronger for utilities than for unregulated firms. See Fred M. Westfield, "Regulation and Conspiracy," *American Economic Review*, Vol. 55 (June 1965), pp. 424–43.

47. Recent moves by large systems to buy abroad, as well as the more passive ("turnkey") purchasing approaches of smaller utilities, are the basis for this suggestion. For some equipment items and coordination arrangements, a coordinated multisystem planning unit would act as a single buyer.

To reach a policy position on competition as a consideration in merger policy one has to make an intuitive judgment. My own is pro-competition, but I am not willing to trade off tangible, scale-related performance effects against the extra competition that might be gained by limiting utility systems to suboptimal scales. The uncertainty of gains from the kinds of competition possible in the power industry, the fact that competition is most effective when it occurs among technically efficient rivals, and the feasibility of attaining network optimization with an industry that is substantially less concentrated nationally than are most major industries call for aiming at the largest number of power systems consistent with a rational power supply from the standpoint of technical efficiency.[48]

Where workable, coordination and wholesaling would be preferable to merger. Merger might be encouraged only in cases where a reasonable effort to achieve coordination by regulators and industry fails to produce adequate results.

TAKING ADVANTAGE OF EXISTING PATTERNS. Scope should be allowed for industry initiative in formulating a workable merger policy. This is a matter of recognizing not only that existing power pools and interconnection patterns have a "going concern" value, but also that the state of readiness for merger is relevant in the timing of a major consolidation movement. Consolidation requires considerable internal reorganization, personnel reassignment, job eliminations, revisions of rate schedules, and the like. Firms differ in the adjustments that must be made and in their readiness to make them.[49] For example, chief executives close to retirement age have less to lose from amalgamation than younger men. Firms that have a satisfactory relationship with the same labor unions and a rapid load growth can solve labor problems connected with amalgamation more easily than firms with demand that is growing more slowly or with more complex or less harmonious labor relations. One advantage of the existing policy, which relies on individual corporate initiative, is that these considerations influence the sequence and pace of merger, and problems of corporate indigestion are less difficult. A disadvantage of total reliance on, and accommodation to, industry initiative is that com-

48. For those who are willing to trade scale-related performance for larger numbers, an appropriately less concentrated benchmark could of course be used.

49. For an informative discussion of some of the considerations involved, see the candid remarks of Sherman R. Knapp, chairman of Northeast Utilities and a key figure in the formation of that holding company, in an interview with *Electrical World* (March 25, 1968), pp. 22 ff.

patibility problems are raised through the long-term effects of mergers on the emerging industry organization. Some firms would be reluctant to merge for a long time; others, when ready to merge, would be limited to second-best alternatives or worse as a result of prior merger by others. A sensible merger policy would combine receptiveness and flexibility with other, more positive ingredients aimed at encouraging an orderly sequence of merger activity consistent with the end-goal industry organization.

OPPORTUNITIES FOR COMPARISON. Although the competitive processes of the free market cannot themselves be duplicated in the public utility field, conceivably the improvement in performance induced by competition could be obtained by some other means. Among the various regulatory and management instruments used for that purpose—rate control, for example—comparison ranks high in potential usefulness.

In idealized free-market form, competition serves as an instrument of search for efficient policies and practices and as a system of control that suggests when search should be initiated. Reward and penalty are closely geared to success in reading competitive market signals, but in the real world market signals convey limited information. A rival's success in charging a low price convinces management that its own costs are probably too high, but it may not know why unless it has access to good industrial espionage.

In the public utility field, rewards and penalties are not nearly as large as in the free market—probably because that is impossible. Regulated utilities, however, have a significant advantage over unregulated industry in the sharing of technical knowledge. The competitive penalty for disclosing information is low, and in any event the regulatory process regularly forces disclosure of detailed data. The rich, open flow of technical, cost, and performance information among electric utilities is in marked contrast to the secrecy among unregulated competitors and provides an opportunity for comparison that is unusual in extremely concentrated markets, regulated or not.

Comparison as a mechanism for search and control works best when the units involved are comparable. American Electric Power would not learn much about thermal efficiency from a medium-sized utility using smaller generating units with a 5 to 10 percent higher heat rate. Comparison of technical efficiency and progress requires the existence of several comparable utilities. This suggests an important advantage to a larger number of systems subject to the constraint that none should be so small as to forgo significant economies of scale under a feasible program of coordination.

EFFECT ON DISTRIBUTION. One effect of consolidating vertically integrated utilities is to increase concentration at the distribution stage. This effect is difficult to evaluate because of the potpourri of issues involved and the tendency for even the best thinking about them to be highly nebulous and necessarily speculative.

The importance of the effect depends in part on whether encouragement by the regulatory agency of specialization in distribution is feasible. The more small to medium-sized systems choose the purchase route rather than consolidation, the larger the number of surviving distributors for any given degree of bulk-power concentration. The most relevant alternative to consolidation is wholesale-retail specialization. The impact of present patterns of wholesale rate regulation deserves study from this point of view. Merger policy is probably not a very good instrument for encouraging wholesale-retail specialization except for the side effect of limiting bulk-power concentration to some kind of end-goal upper bound. Requiring divestiture of distribution property as a condition of approval is effectively equivalent to prohibiting merger.

The effects of merger on technical efficiency and progress in distribution are probably neutral, since the weight of available evidence suggests that the efficiency of distribution systems is primarily dependent on consumption per customer and load density rather than on the size of the franchised territory.

Many of the issues relating to the effect of consolidation on distribution fall under rules already discussed, chiefly those of competition and comparison. The number of efficiently large distributors[50] available for comparison is high, and concentration of major bulk-power systems at the thirty-planning-unit level would change the number very little. The competitive effect of merger might be more important, since the rivalry mainly affected—for retail industrial customers—is frequently in regional markets.

The advantages of local service (or alternatively of service by a large organization) are occasionally cited in policy debate without being spelled out. One can picture small distributors as flexible, responsive to

50. There are currently over 3,500 distributors in the industry, several hundred of which are investor owned. The number that are efficiently large is at least in the hundreds.

customers' complaints and needs, and likely to explore various approaches to special types of improvements in service quality, such as
beautification. From this point of view, the large distributor is pictured as
a sluggish, unimaginative monolith. Or a very different set of images can
be constructed; for example, by stressing the importance of sharing
scarce managerial resources—statesmanlike, farseeing corporation presi-
dents—in a concentrated industry in which small firms are characterized
as having backward management. These tenuous arguments would probably weigh rather lightly in any serious analysis. In any event, policy
makers should hear the arguments, take appropriate readings, and treat
departures from benchmark structure as a conscious sacrifice of technical
efficiency and progress for some other objective.

Aspects of a Merger Policy

Some broad aspects of a merger policy begin to emerge when the six
considerations just reviewed are combined with the basic structural implications of scale economies. Flexibility, compatibility with final goals,
competition where no clear sacrifice of technical efficiency is involved,
recognition of "going-concern" advantages, better opportunities to make
interutility comparisons for improved performance, and efficient distribution are all compatible with the notion of a twenty- to thirty-planning-
unit structure as a working standard. In the aggregate, they enhance the
case for aiming at a number of major thermal bulk-power planning units
closer to thirty than to twenty, possibly in addition to the federal hydro
marketing agencies and such state entities as the Power Authority of the
State of New York. If, on future examination, a thirty-unit industry
should prove insufficiently concentrated to achieve technical efficiency,
additional mergers could then be approved. Since a prodigious amount
of merging may be necessary to arrive at a thirty-unit industry, little will
be lost if more concentration later proves desirable. Decisions to create
systems larger than this benchmark size should be deferred until more
experience is obtained with publicly encouraged power pooling.

A governmentally imposed master plan is neither likely nor desirable.
Industry initiatives should be encouraged if they are broadly consistent
with the desired pattern of size, balance, territorial contiguity, and geographical and historical factors that favor particular groupings of utilities. Many merger and coordination combinations are consistent with the

final goal, and it makes sense to give industry initiative this much range. At the same time, mergers should be blocked if they violate these broad criteria. Examples would be major acquisitions by the largest utilities or mergers that isolate utilities or groups of utilities so that the latter must either remain inefficiently small or later merge into a system that is already large enough. Knowing which mergers are likely to be consistent with the desired ultimate industry organization calls for advance exploration of possible patterns of combination and coordination among contiguous utilities. Advance planning should not lead to hardened decisions but should be used to provide perspective and relevant questions when a particular merger proposal requires regulatory action. The encouragement of merger of balky systems is a delicate task. The willingness of the electric utilities to merge in the 1950s and 1960s suggests that merger initiatives are sensitive to relatively subtle changes in stated public policy. Overzealous encouragement can adversely affect the terms of settlement and can deter utilities from trying to make coordination work efficiently before resorting to merger. Ideally, a good-faith effort to coordinate or to adopt wholesale-retail specialization should be encouraged first. If in some areas the industry is too unconcentrated for effective planning units to function after a few years' trial, regulators should indicate that merger requests from the systems involved will be received sympathetically.

What is feasible under present law is a matter for study by lawyers and, in any event, depends in part on the unfolding of precedent as the consolidation movement proceeds. Nevertheless, a broad interpretation of existing statutes coupled with an activist stance by regulators could go far in directing policy toward the general approach suggested here. The actions of regulators to date follow this approach in part but differ in two important respects. First, the situation that would obtain after merger has generally been compared with the status quo rather than with alternative merger patterns. Second, regulators have not encouraged, at least visibly, consolidations of utilities that have been slow to take the initiative despite obvious potential gains from merger and lack of evidence that adequate progress toward coordination is forthcoming.

Given the enormous resources at stake in bulk-power networks, the public has a right to expect a merger policy that is well thought out in terms of its ultimate implications for industry organization and performance and that will bring about an orderly transition without unnecessary delay.

CHAPTER FOUR

The Competitive Margin in Communications

William G. Shepherd

THE INDUSTRY that provides point-to-point communications services has several unique characteristics that profoundly affect its regulation and innovative performance. Monoply conditions are more closely approximated, both because intermodal competition is peripheral and because the design and supply of equipment are vertically integrated with the common carriers. Public regulation has set entry barriers, but much of its effort to constrain profits and prices has been informal and off the record. The innovative activity within this setting has three significant characteristics: (1) it has ranged from basic to applied science and from rivalrous to total monopoly conditions; (2) it reflects decisions among possible new-product and cost-reduction innovations, and among varying standards of service reliability; and (3) it involves significant interdependence within the basic network.

A large share of the invention and innovation has been performed via coordinated choice within the Bell System—in its Bell Telephone Laboratories, Western Electric Company, and more than twenty operating subsidiaries. Such a complex, internal process is not easy to measure and evaluate objectively, and its responses to—more precisely, its interactions with—regulatory constraints are still more difficult to establish. These problems are compounded by the depth of the commercial interests at stake for both the carriers and potential entrants. The carriers have special incentives to select innovations, to invoke regulatory procedures, and to control the flow of technological information so as to minimize the probability of new entry into any of their actual or desired markets.

New trends in demand and technology have suggested that several parts of the sector may be increasingly amenable to a more, or even a fully, competitive structure. These parts include particularly the large-scale transmission of data, domestic satellites as an alternative to land-based transmission, and the supply of equipment. On some of these areas, there are objective data about the technological trade-offs, but on many others —and on the whole process of technological change in the sector—little is publicly available for scientific analysis. Consequently, it has so far proven impossible to test whether a variety of regulatory controls on entry and prices should be revised or removed, or indeed how regulation may actually have affected innovation. Since the main source of technological information (the Bell System) is also an interested party, its choice of issues, data, and conclusions needs to be subjected to independent verification.

This chapter presents seven hypotheses about the relation of regulation to innovation in the sector, together with available evidence about their validity. Since the hypotheses derive primarily from the distinctive features of the sector, these features need first to be reviewed in more detail. As a brief, selective background summary, the first section discusses the setting of markets, innovation processes, and trends. The section that follows reviews several aspects of regulation and of other public policies that have impinged on the sector. The next section presents the seven hypotheses and notes points for further research. A final section lists several policy implications.

Several basic points of analysis, which help to keep the following details in perspective, should first be noted. In general, incentives to innovate are greater under competition than under monopoly if the benefits of innovation are appropriable.[1] Since appropriability appears to apply to communications innovation at least as much as to innovations in other

1. Kenneth J. Arrow, "Economic Welfare and the Allocation of Resources for Invention," in *The Rate and Direction of Inventive Activity: Economic and Social Factors,* a Conference of the Universities—National Bureau Committee for Economic Research and the Committee on Economic Growth of the Social Science Research Council (Princeton University Press for the National Bureau of Economic Research, 1962), pp. 609–25. See also William Fellner, "The Influence of Market Structure on Technological Progress," in American Economic Association, *Readings in Industrial Organization and Public Policy* (Irwin, 1958), pp. 277–96; Dan Usher, "The Welfare Economics of Invention," *Economica,* Vol. 31 (August 1964), pp. 279–87; and Frederic M. Scherer, "Research and Development Resource Allocation under Rivalry," *Quarterly Journal of Economics,* Vol. 81 (August 1967), pp. 359–94. Monopoly elements may include market shares and entry barriers, both of which are highly relevant to conditions in the sector under review here.

sectors, the rate of such innovation is expected to be reduced wherever regulation creates or permits unnecessary monopoly without taking specific steps to offset this effect. Regulation on a cost-plus-profit basis reduces constraints on all costs, including those for inventive activity. Consequently, the incentives of a regulated carrier as a monopolist are to retard the use of its inventions in actual innovations, although to protect its monopoly position it may engage in more inventive activity than an equivalent unregulated carrier.

The decisions governing technological change in this sector are closely analogous to those of portfolio selection by investors.[2] The choices are among a variety of expenditure sets for prospective innovations, including new services, process changes, spatial configurations, and reliability standards. Each set offers differing return and risk outcomes. The attractiveness of a choice to the firm is determined by both the expected rate of return and the degree of risk. An external regulatory constraint on the rate of return of a utility firm lowers the potential profitability of high-risk ventures, thereby reducing the incentive to take risks. As a result, the constrained utility will alter its selection of technical changes away from high-risk, high-return innovations. In certain cases, it will have incentives to undertake innovations with specific low (or negative) returns, if they greatly diminish risk for the system as a whole. This will cause innovative expenditures to be greater in situations where there is rivalry or potential new entry. If a newcomer is the leader in an innovation, the carrier will respond rapidly and endeavor to reduce the cross-elasticity of demand by using regulatory devices to exclude the newcomer. The faster the response and the greater the reduction in cross-elasticity, the smaller will be the gain to the newcomer.[3] Extreme cases of high cross-elasticity and rapid response may induce specific innovation expenditures by the carrier to go beyond the efficient margin (as defined for that particular innovation). Such a response is normal for all monopolists; regulation

2. This is evident in the discussion by Richard R. Nelson, "The Link Between Science and Invention: The Case of the Transistor," in *Rate and Direction of Inventive Activity*, pp. 568–75. On the analysis of portfolio selection generally, see James Tobin, "Liquidity Preference as Behavior Towards Risk," in Donald D. Hester and James Tobin (eds.), *Risk Aversion and Portfolio Choice* (Wiley, 1967), and Harry Markowitz, *Portfolio Selection: Efficient Diversification of Investments* (Wiley, 1959).

3. M. I. Kamien and N. L. Schwartz, "Market Structure, Rivals' Response, and the Firm's Rate of Product Improvement" (Carnegie-Mellon University, 1969; mimeographed). As Kamien and Schwartz note, a high degree of entry may tend to inhibit innovation under certain conditions.

may simply alter the composition of the innovation set and increase the scope for excessive response.

More generally, by controlling structure and the conditions of entry, regulation is able to accentuate, remain neutral to, or offset the restrictive effect of carrier monopoly on innovation. By constraining profit rates, regulation will tend to induce a lesser average degree of risk taking in innovative activity.

The Sector

The primary components of the point-to-point telecommunications system are the switching subsystem, the transmission equipment, and the terminal devices. Switching is the process of selecting a local or intercity message path and involves electromechanical and electronic equipment. Intercity transmission is by cable, land-based microwave, and terrestrial satellite at selected levels of message-transmission quality. Terminal devices are of great and growing variety, and their interconnection with the system must provide safeguards against damage to the system's functioning. A key variable in choosing technologies in all parts of the telecommunications system is the standard of reliability, a probabilistic criterion which encompasses the determination of both optimum quality of service and optimum capacity during peak loads. The decision to alter technology or service offerings in any of these dimensions must often embrace complex interactions within the system. By the same token, the innovating carrier commonly has a significant degree of choice among technological modes, service offerings, and spatial configurations. This latitude is enhanced if research and equipment design are vertically integrated with the provision of services, as they are in the major communications carriers.

The telegraph was first developed in the 1840s, the telephone in the 1870s.[4] The American Telephone and Telegraph Company's (AT&T)

4. For an outline of the development of the industry and public policies toward it, see Charles F. Phillips, Jr., *The Economics of Regulation* (rev. ed., Irwin, 1969), Chap. 17. See also the landmark study by the then-new U.S. Federal Communications Commission, *Investigation of the Telephone Industry in the United States,* H. Doc. 340, 76 Cong. 1 sess. (1939) (hereafter referred to as FCC 1939 *Investigation*). A review of recent conditions is provided by the *Final Report, President's Task Force on Communications Policy,* prepared during 1967 and 1968 (Government Printing Office, 1969).

intercity linkage of local exchanges by wire was well advanced by 1900. Cable technology has undergone continuing improvements, and in the 1950s land-based microwave transmission (first demonstrated in 1915) assumed a major share of intercity traffic. The emergence of satellite transmission in the 1960s has been paralleled by further advances in land-based alternatives for high-density routes, including pulse code modulation, lasers, and wave guides.[5] Switching technology has progressed through electromechanical to electronic exchanges. Although local dial service was provided later in the United States than in certain European systems, long-distance dialing did not so lag and is now virtually complete. Terminal equipment types have proliferated in the 1960s, as computer time sharing has accelerated the growth in data transmission. Recent service innovations have been primarily in private-line and large-scale data transmission services.

Generally, the United States has offered exceptionally favorable conditions for innovation in communications. Its great size, scattered large cities, and extensive commercial use of telecommunications have all favored the creation of high-density trunk routes that can exploit the major potential advances in transmission technology. These have been the main cost-reducing innovations in the last forty years. Also, at least a large portion of demand is probably income elastic. Rising incomes have brought about high and rising levels of demand for basic and optional services, so that many other innovations have been relatively easy to prepare, test, and carry out.[6]

From the outset the Bell System has controlled more than 80 percent of local exchange capacity and virtually all long-lines capacity, and its Western Electric Company subsidiary has supplied nearly all equipment since 1881. The Bell System is the country's largest single monopoly, and in 1968 Western Electric ranked twelfth in sales among all U.S. industrial corporations. Bell Telephone Laboratories, established in 1924 to perform research and product development, is a leading private industrial research group.

The Bell System has rejected requests by users to gain direct interconnection of non-Bell apparatus with its system (interconnection with other telephone systems has been standardized since 1912). Accordingly it

5. For an excellent summary of present and prospective changes, see Leland L. Johnson, "New Technology: Its Effect on Use and Management of the Radio Spectrum," *Washington University Law Quarterly*, Vol. 1967 (Fall 1967), pp. 521–42.
6. Leland L. Johnson has been especially helpful on these points.

Table 4-1. Value-Added and Value of Shipments of Communications Equipment and Related Products, 1966

In millions of dollars

S.I.C.C.[a]	Industry	Value-added	Value of shipments
3661	Telephone and telegraph apparatus	1,432	2,467
3662	Radio and TV communications equipment	4,855	7,563
367	Electronic components	4,223	6,644
365	Radio and TV receiving equipment	1,841	4,336

Source: U.S. Bureau of the Census, *Annual Survey of Manufactures: 1966* (1969), pp. 44, 46.
a. Standard industrial classification code.

has, until recently, controlled the market for terminal equipment, provided by Western Electric, as well as for other equipment in the system. Yet alternative supply capabilities for such equipment have existed for many years, both in the United States and abroad.

The scope of the communications sector is indicated very approximately in Tables 4-1 and 4-2. Although some firms (such as the Bell System) are integrated through the three partially distinct stages of research and development, equipment production, and system operation,

Table 4-2. Domestic Revenues from Communications Services, by Type of Carrier, 1966

In millions of dollars

Carrier	Revenue
Common carriers	
Telephone: Bell System	12,419[a]
Telephone: Independents	1,734
Telegraph	319
Broadcasting	3,075
Post Office	4,784
Total	22,331
All other carriers (estimated)[b]	15,000–25,000

Source: U.S. Bureau of the Census, *Statistical Abstract of the United States, 1968*, pp. 494, 497, 498, 501, 505.
a. The Bell System breaks down as follows:

Local	6,517
Toll	5,378
Other	524

b. The estimate is based on the ratio of carriers' equipment purchases to total equipment sales. To reflect its admitted crudity, it is presented as a range of probable values.

many firms are not. Attempts at regulation by the Federal Communications Commission (FCC) and state commissions are mainly at the third, equipment-using level.[7] Increasingly the FCC has been drawn as mediator into major corporate struggles, some of them involving very high stakes. At the equipment-producing level, federal agencies have assumed a large role since the early 1950s via their large volume of purchases and payments for research and development (R&D).

Equipment producers can be divided into two distinctive categories: group 1, those owned by regulated carriers, such as the Bell System and General Telephone and Electronics; and group 2, all other equipment-making firms, large and small. Group 2 firms generally innovate and market their equipment under competitive, or at least rivalrous, conditions, although there are tendencies toward "tight" rather than "loose" oligopoly and Western Electric is excluded from competing in these markets. There are many buyers of broadcasting equipment (networks and individual stations), even though one buyer is owned by a major producer—Radio Corporation of America (RCA) owns the National Broadcasting Company—and both RCA and International Telephone and Telegraph Corporation (ITT) have a variety of carrier operations abroad. Since the networks' earnings are not regulated, rate-base preference is not a factor in their purchasing policies. Some items are affected by international competition. Market structure is particularly competitive in the smaller, peripheral products, such as terminal attachments for data transmission.

There are exceptions. Federal procurement of communications equipment and services generally takes place in the absence of price competition. Also, rising federal support financed about two-thirds of all communications R&D by the end of the 1960s. The contracting procedure for federally supported R&D has diverged from standard patterns of price competition in decentralized markets. Yet in both areas—procurement and R&D contracting—there have been strong elements of nonprice competition, and some efforts have been made to avoid excessive reliance on single sources. Moreover, the military and civilian markets do not overlap extensively, so that the civilian spillover of both benefits

7. Equipment prices of the Bell System and General Telephone and Electronics Corporation (the largest independent) are informally regulated. The equipment-producing subsidiaries submit rate-of-return calculations on the main categories of operations to demonstrate that excessive profits are not being concealed in equipment prices. See Phillips, *Economics of Regulation*.

and anticompetitive effects from military activities by communications firms has been relatively limited.

Conditions for equipment producers owned by regulated carriers (group 1) depart more distinctly from competitive market patterns in four main respects.

First, the R&D performer and product developer has a monopoly position and is vertically integrated. For many services, the ultimate users have no alternatives for comparison, either in specific equipment or in the quality and variety of services. The process of innovation is analogous to portfolio selection, as noted in the introductory section.[8] The Bell System does apply a sequential system of project evaluation that includes calculations of costs and net returns at various stages as an invention proceeds to systemwide application. Indeed, the three stages of invention, innovation, and imitation are highly commingled and interacting within the integrated system.[9] Yet the system's private criteria do not ensure that the socially optimal choices will be made. Unless they are constrained or induced in other ways, invention and innovation by the monopolist are likely to be less than would be socially optimum.[10]

Second, public regulation sets the borders of the utility's franchised monopoly markets and formally constrains the firm's rate of return on carrier service (and indirectly on equipment supply). Regulation does not directly influence inventive or innovative activity, but its possible side effects on them are many (see the section entitled "Seven Hypotheses" below). And the carriers' aversion to service failures, the most sensitive point for public disapproval, is reinforced by the regulators' aversion; both groups prefer high standards of service quality and reliability.

Third, some R&D is financed from carrier revenues. Standard regula-

8. See especially Nelson, "The Link Between Science and Invention," pp. 568–75, and Thomas A. Marschak, "Strategy and Organization in a System Development Project," pp. 509–48, both in *Rate and Direction of Inventive Activity*.

9. See "Vertical Integration in the Bell System: A Systems Approach to Technological and Economic Imperatives of the Telephone Network," staff papers submitted to the President's Task Force on Communications Policy (reproduced by the Clearinghouse for Federal Scientific and Technical Information, June 1969), for an appraisal showing that the interrelations are extensive.

10. See Fellner, "Influence of Market Structure," Arrow, "Economic Welfare," Usher, "Welfare Economics of Invention," Scherer, "Research and Development Resource Allocation," and Kamien and Schwartz, "Effects of Market Structure."

tory criteria, based on cost plus profit, tend to reduce the carrier's incentive to limit costs. Regulation, therefore, is likely to induce higher spending on R&D than the unconstrained profit-maximizing monopolist would choose, whether or not the results of it come to be used in actual innovations. The monopolist's incentive in blanketing R&D possibilities will be strengthened by the lowered opportunity cost of R&D expenditure. The system will be in a position to respond more quickly to entry-opening innovations by outsiders or to make technological changes that preclude such entry altogether.

Fourth, the "natural monopoly" portions of the system (primarily in local and intercity switching, but also in some aspects of equipment design and production) are mixed with activities whose technology is increasingly likely to permit a rivalrous or fully competitive structure of supply. The precise extent and configuration of these conditions are debatable, both because estimates of them are imperfect and because their policy lessons involve large commercial interests. Suffice it to say here that such competitive possibilities will induce carrier responses that contrast with monopolists' innovating behavior in the sector's noncompetitive areas (this is discussed more fully below, under "Seven Hypotheses"). Generally, innovations providing private benefits from increasing the technological exclusivity of the system will be favored over more neutral ones.

To summarize, innovation by group 1 firms is likely to range from inadequacy in protected areas to rapidity in competitive ones, even though R&D resources are greater than an unconstrained monopolist would expend. Innovations will tend toward system exclusivity, and there will be a strong preference for vertical integration and a reluctance to clarify its alternative costs and benefits.[11]

Regulatory Issues and Measures

Regulation and antitrust are the two main public policies that may have affected innovation in the sector. Common-carrier telephone services have long been recognized as having "natural monopoly" features, which is why a franchised monopoly status has been granted to certain

11. While it is not clear that regulation affects these exclusivity preferences, one task for regulators will be to anticipate such preferences and offset their effects.

telephone systems for nearly a century.[12] A system must stand ready to provide connections among its subscribers, and service standards may require unified control over all components and possibly self-supply of key apparatus. Hence, the telephone carriers (particularly the Bell System) have argued that there will be large costs from reducing system "integrity," and the Bell System has resisted interconnection with other systems and the attachment of terminal devices not produced and serviced by itself.[13]

If the technology of telecommunications prescribed a natural monopoly, total and integrated control by one system under appropriate public constraints might be economically justified. Yet the technological margins between "necessary" monopoly and possible competition have been in a state of flux, particularly in the last two decades, and the newer "optional" services have tended to be both more lucrative and more capable of supporting a more competitive market structure.[14] Where these new markets are not already carrier-controlled (as in optional handset features), the common carriers have had an incentive to capture the entire

12. See Phillips, *Economics of Regulation*. For a critical review of the "natural monopoly" basis for regulation, with special reference to telecommunications, see Richard A. Posner, "Natural Monopoly and Its Regulation," *Stanford Law Review*, Vol. 21 (February 1969), pp. 548–643; also the skeptical review by James C. Bonbright in his *Principles of Public Utility Rates* (Columbia University Press, 1961). Posner's summary is valuable, though he does not establish that the net costs of regulation clearly exceed those of the feasible alternatives; see my "Regulation and Its Alternatives," *Stanford Law Review*, Vol. 22 (February 1970), pp. 529–39. On the slow early steps toward telephone regulation, see the FCC 1939 *Investigation*, and Phillips, *Economics of Regulation*, Chap. 17 and sources cited there.

13. Full interconnection with independent telephone companies for long-distance service has been provided since the so-called Kingsbury Commitment of 1913; see Phillips, *Economics of Regulation*, pp. 655–56.

14. This appears to be a possibility for many of the newer services, such as data transmission and private exchanges for large customers. It is also a possibility for basic intercity microwave (and very possibly satellite) transmission activities, where routes are duplicated. It may apply also to the manufacture of much telecommunications equipment, even though the Bell System now buys virtually all of its equipment from its own Western Electric subsidiary, because many American and international firms have (or could develop) capabilities of supply. See particularly the record in the Carterfone, Microwave Communications, Inc., and computer time-sharing hearings by the FCC in the late 1960s (FCC Dockets 16942 and 17073, 16509–16519, and 16979, respectively), and Phillips, *Economics of Regulation*, Chap. 17. These possibilities are mentioned here only as illustrations of the variety of areas that may be involved.

markets for these services by persuading the regulators to classify them as exclusively "carrier" activities. Failing that, the carriers may have been able to offer newer services at prices that independent firms, lacking the carriers' established earnings base, cannot meet.[15] Meanwhile, independents have been arguing that the carriers (again, primarily the Bell System) could permit terminal attachments and interconnection with private systems much more freely than they do without significantly reducing system integrity and service quality. These and related changes could greatly enlarge the scope for competition, or at least rivalry.

There is thus a large public stake in the technological trade-offs: in the degree of alternative costs and gains from reducing monopoly barriers, and conversely in the losses of efficiency and innovation that are likely to be caused by overextending the franchise for monopoly. In all of the variables in question, one needs to consider alternative gradations of change rather than simple comparisons between the status quo and extreme alternatives. For example, assertions about technological "requirements" or claims that vertical integration yields great economies of coordination do not suffice as a basis for public policy, especially where they come from carriers with an interest in excluding competition. Not only may the carriers' own internal perception of the trade-offs be conditioned by their interest in increasing the reach of their control, but in addition the very nature of the innovations they adopt may tend to be selected so as to maximize the exclusivity of their systems. This will indeed be so if the carrier is behaving rationally and if it has a significant degree of choice. The more integrated the system, the more scope there may be for selection among innovations, and the harder it is, even for alert regulators, to perceive, anticipate, and offset the bias. As noted earlier, reliability of service frequently plays a key role in the carriers' strategy for innovation and profit maximizing.

The strong incentives of carriers to increase the reliability and quality of service may shape the whole direction of technology. Regulators share

15. The modern classic on this is Harvey Averch and Leland L. Johnson, "Behavior of the Firm under Regulatory Constraint," *American Economic Review*, Vol. 52 (December 1962), pp. 1052–69. In the extreme case, "Our model suggests that apprehension about the nature of competition in the industry is justified since a common carrier, regulated as described above, would (under certain conditions) have an incentive to operate at a loss in competitive markets and to shift the financial burden to its other services" (p. 1065). To this extent, regulation may accentuate the price discrimination that a monopolist would use; this is discussed further in the next section.

some of these inducements (for example, to avoid embarrassing stoppages in service), so they must be particularly alert to the likelihood that the degree of reliability and exclusivity will be extended beyond the technological optimum.

Complex regulatory problems have existed since the 1880s, long before there was much formal regulation, but the public stake in them is now higher. The standard regulatory process in this country is concerned basically with two areas—utility behavior and market structure. The controls on behavior usually aim to restrict profits to "fair" levels and rate structures to "just and reasonable" patterns, whatever those adjectives may mean. Profitability has two prime elements, rate of return and risk, both of which should be considered by regulators. Most state regulatory hearings have taken a narrow and mechanical view of rates of return, with little analysis of risk.[16] Risk (and therefore profitability) is also directly affected by regulatory controls on structure, which can shield the carrier absolutely or can tend to foster competition; that is, the regulators make or prevent much of the risk. Indeed, regulation's protective role for many carriers is probably more important than its efforts to restrict rates of return. The argument has often been made in recent years that structural regulation does tend to be excessively protective.[17] Regulation in practice often appears ideally suited to the carriers' long-run interests, since it reduces their risks and hardens and extends their market control. This outcome may reflect deliberate policy decisions; but often it is caused instead by regulatory incompetence and inadvertence, which in turn are often results of gross underfunding. Frequently the carrier's original franchise is simply continued, even though new technological conditions would permit competition. Or regulatory procedures may be manipulated to capture new and potentially competitive services; the carrier simply issues a tariff on a new service, even if it has been developed by others. By approving the tariff (a move difficult to avoid without seeming to stand in the way of service innovation), the regulatory agency formally, and often irreversibly, excludes further competition.

16. This may be inevitable in state regulation of the more than twenty Bell operating companies, for the genuine risk attaches primarily to the entire Bell System rather than to the state-level operations of its component companies. Since most economic risk is therefore outside the reach of state regulation, the actual preoccupation with narrow rate-of-return issues may be quite inevitable and all the more peripheral.

17. See Posner, "Natural Monopoly," and sources cited there.

Structural issues such as these have recently come to the fore in the communications sector, as will be seen below. Yet only limited efforts have been made to calculate the net benefits and costs of structural changes. And it is commonly in the carriers' interest to avoid such appraisals of alternatives.

In contrast to these complexities, early public policies toward telegraph and telephone systems were relatively simple, permitting virtual monopolization of each of the modes and vertical integration of equipment production and carrier activities in telephone systems.[18] The threat of antitrust action did produce the agreement in 1913[19] that limited Bell's acquisition of independent telephone companies and provided for their interconnection with the Bell System. Western Union was also separated permanently from AT&T. In short, early public policies had some *structural* impact, though it was largely peripheral and lacking in objective research into the technological trade-offs. These moves ratified the basic monopoly positions without providing for regulation of behavior. Although some states had created regulatory commissions with this power by the 1920s, their authority was almost always used sparingly and it could not in any case affect interstate activities.[20]

By the 1930s, use of the telegraph was clearly on the decline, eroding Western Union's monopoly position and leaving it a relatively well-defined and specialized problem for public policy, though far from an easy one.[21] By contrast, AT&T had become (and has remained) the largest monopoly in the economy. During the years 1935–39, the newly created Federal Communications Commission examined the situation in detail, though again without systematically researching the technological alternatives.[22] The FCC found many opportunities for abuse in the system, particularly in the vertical integration of equipment production by Western Electric, but it concluded that its own close surveillance could prevent such abuses and provide, for the first time, "continuous and efficient" regulation of the industry, or at least of its interstate aspects.[23]

In fact, *intra*state telephone activities have remained under the widely

18. See the FCC 1939 *Investigation*.
19. See note 13.
20. See Phillips, *Economics of Regulation*.
21. See the FCC study in the 1960s of the future of Western Union, Docket 14650; see also the 1967–68 *Final Report* of the President's Task Force on Communications Policy.
22. FCC 1939 *Investigation*.
23. Ibid., p. 602.

varying individual state commissions, some of which are motivated and equipped to provide close regulation and some of which are not. Most state commissions have meager resources in relation to their responsibilities and problems and to the carriers' opposing reserves of talent. Generally they devote a large part of these meager resources to housekeeping matters. Moreover, thirty-six states, with about half of all Bell telephones, are served by *multi*state Bell operating firms (such as Southwestern Bell Telephone which provides Missouri, Kansas, Arkansas, Oklahoma, and Texas a total of 9 million telephones). The scope for effective state regulation has been reduced as the distinction between interstate and intrastate operations has become increasingly blurred.[24]

Many commissions have been genuinely passive to carrier policies on most issues, particularly on rate structure, service offerings, and investment decisions. Moreover, even the few relatively tough commissions (in California, New York, Wisconsin, and Michigan) have lacked the staff and the perspective to go much beyond a general effort to limit aggregate rates of return.

The FCC, on its part, relied primarily on "continuous surveillance" from 1939 to 1965. Apart from the specific cases mentioned in the next paragraph, until the major investigation began in 1965 the FCC made no effort to submit to public hearings the industry's basic structure or the economic activities going on within it (rates of return, rate structure, interconnection and attachment policies, depreciation policies, equipment purchasing, and so on). In relying on negotiation, the FCC gave up the leverage and initiative that public hearings provide in dealing with powerful interests and complex issues. And it left both the character and the results of its actions unavailable to outside study. Continuous surveillance has prevented the rigidities and formal posturing that regulatory hearings too often create—an important gain. Moreover, a series of rate changes has emerged from these informal negotiations. Possibly the FCC has significantly influenced those changes; but outside observers are left with neither descriptive data nor normative guidelines for appraising them or for factoring out the FCC's role, if any, in bringing them about.

Public hearings have been held on a series of unavoidable new issues

24. There is no clear technological basis in the pattern of single-state and multiple-state companies; single-state companies range from very large (New York, with 11 million telephones) to very small (Delaware, with only 300,000). The situation appears to be ripe for revision, with perhaps a major extension of FCC jurisdiction. Yet the FCC has made no move to assert this jurisdiction, and indeed it does not have the staff to do so.

affecting the structure of communications markets, particularly spectrum allocation, microwave access and interconnection, the division of data transmission between Western Union and the Bell System, foreign attachments, and satellite transmission. Most of these have been lengthy, perhaps befitting the depth of the problems, but they have tended not to open up large new competitive possibilities.

Meanwhile, since the 1930s, action by other agencies has posed some major issues. In 1949, the Justice Department brought suit under Section 2 of the Sherman Antitrust Act,[25] seeking divestiture of Western Electric from AT&T. In 1956, the case was settled before it reached trial by a consent decree that offered several modest gains but left Western Electric's attachment to AT&T intact.[26] No way was provided for outside suppliers to compete in the sale of equipment to Bell operating companies, and little has been done in this direction since. At the same time, Western Electric agreed to produce only telephone equipment, for sale largely in the Bell System and to government agencies; since then it has not bothered to enter competition in foreign markets, regarding itself as fully occupied with domestic responsibilities.[27] Thus the consent decree operates as a *détente,* which at one stroke eliminates two possibilities for new competition. It leaves intact what is probably the country's largest industrial monopoly (in Bell System supplies, currently about $3 billion annually) and at the same time it excludes Western Electric from competing in other communications and electronic equipment markets, many of them substantial. Despite apparently firm legal ground for reopening the issue, no change now appears likely, although in 1968 the matter was under further review by the Justice Department.[28] The Bell System re-

25. *United States* v. *Western Electric Company, Inc., and American Telephone and Telegraph Company,* Civil Action 17-49, *Complaint,* 1949.

26. The consent decree settling the case was entered Jan. 24, 1956, in the United States District Court of New Jersey. For a criticism of this settlement, see *Report of the Antitrust Subcommittee on the Consent Decree Program of the Department of Justice,* U.S. House of Representatives, 86 Cong. 1 sess. (1959); also John Sheahan, "Integration and Exclusion in the Telephone Equipment Industry," *Quarterly Journal of Economics,* Vol. 70 (May 1956), pp. 249–69.

27. A recent outline of the resulting anomalies in Western Electric sales to other telephone systems is given in "Small Phone Companies Switch On," *Business Week* (May 10, 1969), p. 160.

28. The Communications Policy Task Force did review the issue, but not in depth or with new data, and it did not recommend change. The issue was thereby frozen for nearly two years with no useful result.

mains unique among major utilities in being fully integrated into manufacture of virtually all of its equipment and in funding R&D activity by a charge on the operating companies. Vertical integration in the General Telephone and Electronics Corporation is simply an echo of the Bell arrangement.

In other countries vertical integration has been unusual, and most systems have several suppliers, at least some of which are separate from the operating system. These equipment suppliers have shown many of the collusive and conservative tendencies that tight oligopoly commonly yields. While some of this behavior could have been prevented by more aggressive purchasing strategies by the systems,[29] experience in other countries offers no assurance that vertical integration is superior to well-managed alternative structural arrangements; the trade-offs are matters for detailed and objective study.

In contrast to the abortive effort in 1949 to introduce competition via antitrust action, during the 1950s equipment purchases and R&D support by the Defense Department and the National Aeronautics and Space Administration came to influence strongly the flow of technology in certain areas of communications. Several important innovations in the sector have arisen from military development efforts, publicly funded. Microwave technology resulted in part from research for military purposes during World War II, and the massive growth of booster satellite-tracking technology is largely the result of the efforts of federal agencies.[30]

Seven Hypotheses

The broad effects of regulation on the rates and directions of innovation in communications have been outlined above, but they do not yet lend themselves to systematic research. They are part of complex multi-variate situations, and the objective data needed to factor them out are not available. One must resort instead to evidence on specific markets

29. On the British instance of this, see William G. Shepherd, "Alternatives for Public Expenditure," in Richard E. Caves and Associates, *Britain's Economic Prospects* (Brookings Institution, 1968). There are indications that future purchasing policies by the British system will be more effective: for example, the market-sharing contracting system is being stopped, and foreign sources of supply are being used for the first time.

30. These and related instances are discussed in the next section.

and innovations. This carries a large risk of sample bias, but it does clarify the magnitudes of possible effects. The seven hypotheses that follow cannot yet be proved or disproved econometrically. But the underlying tendencies toward exclusivity and differential response to entry can be tentatively established. Some of the seven hypotheses[31] derive from the core of classical regulated-monopoly carrier conditions; but perhaps the more interesting and important ones arise from the shifting boundary between monopoly and competition, in both equipment manufacture and the provision of services. The first four hypotheses are primarily behavioral; they make assumptions as to how regulation may affect behavior within a fixed structure of markets and regulation. The last three are structural; they outline the interactions that may occur among regulation, technology, and the structure of utility markets over time.

Classical Monopoly Restraint

If a carrier prices and innovates as a protected monopolist, R&D and investment in new technology, as well as output, will be reduced.[32] Innovation will then be smaller in volume, slower, and less varied than it would be under competition, or under the degree of competition that generally prevails in the economy.

Pricing and growth policies in the communications sector do not appear generally to follow the classic monopoly model, either in the manufacture of equipment or in the provision of common-carrier services. Perhaps the exception is innovation which may reflect somewhat greater restrictiveness. Industry critics repeatedly suggest that the adoption of several advances by the Bell System was slowed by its preference as a monopolist for Western Electric equipment. These innovations include automatic switching, unattended-dial central-office equipment for small exchanges, and modern handsets, none of which were adopted by the Bell System until the 1920s.[33] Another hint is in the contrast between

31. For a discussion of several of these points, see Posner, "Natural Monopoly."
32. See, for example, C. Emery Troxel, *The Economics of Public Utilities* (Rinehart, 1947). A primary concern until recently was that utilities, with the added protection of regulation, would be *restrictive* in all activities.
33. FCC 1939 *Investigation,* pp. 218–20, 245–46, 584–85, and Sheahan, "Integration and Exclusion," pp. 262–65. Sheahan notes that the amount of delay on these is somewhat ambiguous. See also the Department of Justice suit to divest Western Electric, Civil Action 17-49, *Complaint,* 1949.

attachments (terminal equipment) for use on the public dial-up network and those for use on private leased lines. The Bell System until 1968[34] permitted only its own terminal apparatus to be used for service on the dial-up public network, although it allowed other attachments to be used on leased lines. This stimulated a rapid proliferation of new devices for use on leased lines, far exceeding the variety of equipment offered by Bell to dial-up users. Moreover, a partial relaxation of the prohibition on "foreign" dial-up attachments in 1968–69 caused a further spurt of innovations by independent producers of all sizes.[35] Bell attachment innovations have been relatively fewer and slower by comparison, further indicating that much potential innovation has been lost precisely because there had been no hope that innovations would gain access to potential markets.

Bell officials have noted significant and persistent differences in innovativeness among the Bell operating companies, even though the operating companies are assessed at the same rate to support Bell Telephone Laboratories and pay the same equipment prices. This suggests, at the least, either that state regulation has little power to appraise, compare, and act to improve innovative performance, or that only some state commissions have been willing and able to try.

X-Efficiency

Studies that estimate the loss of income due to monopoly and to tariff restrictions have generally been based on the assumption that production functions, and therefore cost schedules, would be the same with or without monopoly (or tariffs). Leibenstein argues that the concept of efficiency underlying these studies is too narrow and that a broader concept, which he calls "X-efficiency," would be more significant.[36] He holds that it cannot be assumed that production and cost functions are unaffected by structural factors, such as the presence or absence of competition.

34. See U.S. Department of Justice, "Memorandum of the United States Modifying Its Prior Petition for Partial Suspension" (Dec. 2, 1968), and Federal Communications Commission, "Memorandum Opinion and Order," FCC 68-1234 (Dec. 24, 1968).

35. This relaxation is discussed further under "Exclusivity of Technology," below.

36. See Harvey Leibenstein, "Allocative Efficiency vs. 'X-Efficiency,'" *American Economic Review,* Vol. 56 (June 1966), pp. 392–415; Armen A. Alchian and Reuben A. Kessel, "Competition, Monopoly, and the Pursuit of Money,"

Cost-plus-profit regulation tends to decrease the incentive to make cost-reducing innovations; and it tends also to induce added spending on R&D (as just another cost), even though the research findings may not result in actual innovations. A carrier's R&D and innovation activities under regulation are therefore indeterminate; in specific cases they may diverge on either side of the optimum. The carrier may choose to do only enough R&D to keep abreast of the possibilities for innovation and to assure that it can invoke regulatory procedures in time to prevent entry of competitors, or it may greatly expand R&D, partly as an end in itself.

In any industry X-efficiency is difficult to appraise, and this is particularly true in communications, with its complexities and lack of comparisons. It could be substantial in the most sheltered parts of the Bell System and in others, including Western Union, but there is no hard evidence to confirm or deny this.[37] Bell Telephone Laboratories is widely thought to be managed efficiently, and Western Electric has long provided price comparisons on selected equipment items in order to estab-

---

in *Aspects of Labor Economics,* a Conference of the Universities—National Bureau Committee for Economic Research (Princeton University Press for the National Bureau of Economic Research, 1962); Oliver E. Williamson, "Managerial Discretion and Business Behavior," *American Economic Review,* Vol. 53 (December 1963), pp. 1032–57; and William G. Shepherd and Thomas G. Gies (eds.), *Utility Regulation: New Directions in Theory and Policy* (Random House, 1966).

37. In March 1969 AT&T issued *A Study of Western Electric's Performance,* a report prepared by McKinsey and Company, a management consultant firm. The study is lavish in its praise of Western Electric management, but several curious features seriously reduce its scientific value. The first is the fact that it was published at all; most such studies are confidential. Second, most such studies focus on needed improvements. Instead this study primarily congratulates; only a few items are glossed over as efficient *and* "being improved" (see p. 34). Such prose is likely to have been prepared for the public, not for Bell management. Third, several tangential topics have been spliced in, even though they have little relevance to Western Electric efficiency. For example, the long passages on the benefits of vertical integration (pp. 70–78, 188–90) appear to have been inserted for the wider Bell System aim of making a case for such integration. Fourth, the hard performance data boil down to cost trends, which have some factual interest but very little normative content. Finally, some invalid comparisons are made with equipment delivery periods in Britain (pp. 25–27). In fact the British situation is affected by unusual conditions of extreme and unexpected growth, comparable to the fundamental problems that have recently arisen in New York City. Since McKinsey and Company has also been doing a study in depth of the British Post Office, such shallow comparisons in this context give rise to serious misgivings about its objectivity. The same specious comparison is offered in "Vertical Integration in the Bell System," Staff Paper 5.

lish its superior efficiency.[38] Yet these usually cover only simpler, periph-eral items, such as handsets, rather than high-value system components; and in any case the data are of little or no probative value.[39] As has long been recognized, they do not show what the comparisons would be if potential competitors were producing on a large scale and were exploring new technologies from which they are presently closed off. The prices are also internal transfer prices, rather than market prices. Their patterns may reflect specific pricing strategies by the entire Bell System, which represent a compromise between enhancing price comparisons in compet-itive parts of the company and increasing the rate base in the regulated parts.

The past high technical quality of Bell System service proves little, and objective data on internal efficiency in providing that service is scanty.[40] Western Union has probably displayed massive X-inefficiency, but making such a judgment is complicated by the special problems of retrenchment with which it has had to cope.[41]

Because of a lack of evidence, the question of X-efficiency must re-main open. This is not crucial here, since the problem is tangential to the main topic of this chapter, the innovative performance of the industry.

Rate-Base Preference and Capital Intensity

Because regulation limits the rate of return on investment, a utility may tend to choose a more capital-intensive technology than it otherwise would. For example, it may enlarge the rate base by (1) carrying extra reserve capacity, back-up stocks, and so forth, of a given technology; (2) employing greater capital intensity in the actual design of capital equipment; (3) owning rather than leasing equipment used in producing services; and (4) tolerating collusive pricing of purchased equipment at a level higher than that which aggressive purchasing policies could have achieved.[42]

38. Francis Bello, "The World's Greatest Industrial Laboratory," *Fortune,* Vol. 58 (November 1958), pp. 149 ff.

39. See, among others, the FCC 1939 *Investigation,* pp. 301–21.

40. One must resort to such scattered hints as a report that a "job enrich-ment" program had cut certain staff requirements by nearly half by reducing long-established bureaucratic controls; see *Business Week* (April 19, 1969), pp. 88–89.

41. See the FCC study of Western Union's future cited in note 21.

42. See especially Averch and Johnson, "Behavior of the Firm"; William G.

The net effect of rate-base preference may be difficult to isolate in actual cases. Yet in communications it may be important, although the several indications of it may simply reflect exclusivity preference instead of (or in addition to) rate-base preference.

First is the choice made during the early 1960s between two alternative satellite technologies. Much of the R&D had been performed by the Bell System, which by 1962 was—along with others—strongly advocating a random-orbital system as the best technology. This would have required a large capital investment in about fifty satellites and in complex and expensive tracking stations.[43] The distinctly less capital-intensive alternative was the synchronous-orbit method, with just a few stationary high-capacity, high-altitude satellites and relatively simple, fixed transmitting stations. The synchronous approach prevailed quickly after 1963, but only because it was invented, developed, and pushed through for competitive reasons by an outsider, Hughes Aircraft. By some estimates this occurred at least five years and hundreds of millions of dollars sooner than the carriers would otherwise have achieved it.[44] Of course, the issues are complex, and it cannot actually be proved that the carriers' perceptions of technical trade-offs and their appraisals of alternative costs and benefits reflected rate-base preferences for the medium-altitude, random-orbital system. Yet the degree and timing of the carrier commitment is strongly suggestive. One notes also that it was not the regulators who perceived and forced through the better technology, but a potential entrant.

A second example is the preference of carriers for owning capacity rather than leasing it. Leasing is frequently more efficient than owning, particularly for office space, vehicles, and a variety of equipment.[45] Yet

---

Shepherd, "Regulatory Constraints and Public Utility Investment," *Land Economics,* Vol. 42 (August 1966), pp. 348–54; Fred M. Westfield, "Regulation and Conspiracy," *American Economic Review,* Vol. 55 (June 1965), pp. 424–43; and Posner, "Natural Monopoly."

43. See Lawrence Lessing, "Cinderella in the Sky," *Fortune,* Vol. 76 (October 1967), p. 201, and Charles E. Silberman, "The Little Bird That Casts a Big Shadow," *Fortune,* Vol. 75 (February 1967), pp. 108 ff.

44. Lessing, "Cinderella in the Sky," pp. 198, 201. Unfortunately there are no detailed studies of these savings in the public record.

45. Criteria for the lease-own choice, like those for reliability of service, present a complex and fruitful field for further research by the FCC or other outside groups. The technological trade-offs are currently a matter of internal Bell System concern, even though they determine important policy choices. Without further study, the comments here must be inconclusive.

the Bell System, for one, seldom leases such items. And its eagerness to participate as owner of at least the ground stations for satellite transmission is also suggestive.

Of particular significance is the Bell System's resistance to proposals that it take the role of a leaser of the satellite circuits for international communications. In this role the Bell System would use leased circuits in place of some of its own undersea cable capacity, which is more expensive but which provides a large amount of rate base (and furthermore is produced by Western Electric). A related specific episode occurred in early 1968, in which the laying of TAT-5, a new United States–Europe undersea cable, was finally approved by the FCC (as recommended by the Defense and State departments) despite estimates by the staff of the Task Force on Communications Policy that satellites would be much cheaper and at least as reliable. One facet of the issue was that choosing a cable instead of a satellite would have meant a substantial net increase in the Bell System rate base.[46]

Bell officials have asked in particular what their stake would be as a mere lessor-jobber of transmission, and they have suggested the need for a special rate-base insert to accompany any such large-scale circuit leasing. They ask this despite the obvious fact that their services would be fully covered as allowable costs. The question and suggestion reflect the System's self-image, under regulation, as a holder-manager of rate base rather than as a service provider, with or without rate base, seeking the best among alternative cost and service combinations. (They reflect also a realization that the holder of the capital, as the Bell System has been in the past, has primary control over pricing, innovation, and access by new entrants; this is discussed below under "Exclusivity of Technology.") Actual choices are shaped by a variety of conditions, including technical requirements, the need for system reliability, and so forth.[47] The lease-ownership issue should not be stretched too far until a more thorough study is made.

A third possible example is Bell's choice of depreciation policies—

46. See Posner, "Natural Monopoly," p. 617, note, and pp. 633–34; Dissenting Opinion of Commissioner Nicholas Johnson from letter of Rosel H. Hyde, chairman, FCC, to Richard R. Hough, vice president, AT&T (Feb. 18, 1968), FCC 68-212; and Dissenting Opinion of Commissioner Johnson from "Memorandum Opinion, Order, and Authorization," Files P-C-7022 and S-C-L-40 (May 22, 1968), FCC 68-569.

47. The Bell System does lease some international satellite circuits. The question is whether without rate-base incentives it would lease more of them in the long run after all adjustments were made.

using long service lives which result from basing a major portion of depreciation on physical wear and "evolutionary" obsolescence (such as replacement due to inadequate capacity) rather than true technological obsolescence (arising when the average total cost of installing a new technology is less than the marginal cost of operating with the old). Twenty- to fifty-year equipment lives have been common.[48] Although this reflects past equipment lives—for example, of the large volume of pre-1920 step-by-step central-exchange equipment, which is still in use—it may not hold in the future. Some major innovations, especially in satellite technology, may displace earlier modes quickly. And the Bell System's planned massive changeover to electronic switching between 1965 and 2000, which is now expected to cost $30 billion, could be rendered obsolete long before it is completed if computer technology advances as rapidly as many experts now expect. In contrast to the currently anticipated thirty-five-year changeover period, computer technology is commonly expected, as in the past, to produce a new generation about every ten years.[49] This could create a serious conflict between the efficient treatment of sequential innovation and the traditional equipment policy in the system. At the least, it raises issues of innovation strategy over time and of the relative costs of varying degrees of commitment to any one capital-intensive technology that would repay independent research.

Slow depreciation keeps high the volume of past investment that is eligible for inclusion in the rate base. It tends to place a floor under the degree of capital intensity of innovations permitted to displace existing equipment, thereby creating a ratchet effect which favors increasing capital intensity. That it also tends to slow adoption of innovations which would destroy existing rate base "before its time" is only one tangible, and perhaps avoidable, side effect. But it is one that regulators may need increasingly to guard against, if that is possible.

The peculiar problem of rate base that the Communications Satellite Corporation (Comsat) has faced in recent years deserves mention. Comsat's initial capitalization of $200 million in 1964 was large enough to fund the then-expected random-orbital system. Since the synchronous

48. At present, most central-office equipment is assigned a life of about twenty years, electronic exchanges about forty years, and underground conduits about seventy years. FCC Orders on modification of depreciation rate for Bell System companies (June, September, November, and December 1968).

49. The point is valid even though most of the changeover to electronic switching may be completed by 1985.

approach cut the requirements by perhaps three-fourths, Comsat has carried well over $130 million in excess capital, which it has put mainly into portfolio holdings. Comsat is therefore in the awkward position of *seeking* rate base in a situation where clearly its technology needs much less than its economic incentives would prescribe. At present, its main effort in this direction is to gain ownership and control of the ground stations.

The other carriers' incentives and opportunities for rate-base expansion are normally vastly more numerous, intricate, and hidden. Only a few have been noted here, and they are to some extent ambiguous. They may be regarded as consistent and significant, and they may well be merely the tip of an iceberg. Perhaps they are about as clear as can be expected in the complex situations common in communications. Even so, they still do not yet prove that rate-base preferences have caused important losses in efficiency or innovation, nor do they indicate how regulators might effectively anticipate and prevent such losses.

Innovation Stopped by Market Closure

Particularly in equipment markets, exclusions that reduce competitive pressures usually also reduce the pace and variety of innovation. A doubly costly two-way barrier, such as the 1956 consent decree[50] provided, may mean than some potential innovations are never developed and others that are developed lie unused.

This effect is perhaps the most conjectural of all. Even where it is known that innovations have been forestalled, they are usually hard to evaluate quantitatively, and the number that are never developed at all may be much greater. Barriers to competition can discourage innovative efforts or kill them off altogether, and the communications field is presumably no exception.

There are some tangible indications of both explicit and implicit losses from the consent-decree partitioning off of the Bell System. The Carterfone decision of 1968,[51] which unexpectedly opened up a major part of the attachments market, stimulated a wave of innovation by independent firms in terminal and interfacing equipment, gave rise to many new firms, and widened the margins and variety of technical capabilities.

During the years 1965–69, as a result of the Bell System's allowing

50. For details of the consent decree, see Phillips, *Economics of Regulation,* pp. 671–73.
51. FCC 68-661, June 26, 1968.

non-Bell attachments to leased lines, innovative activity in leased-line attachments differed markedly from such activity in dial-up attachments. One must rely more on general impressions and professional opinion than a scientific assessment would permit. But the innovative efforts in terminal equipment for attaching to leased lines were almost certainly more vigorous and varied than those for dial-up attachments, however they might be measured. This reflects the greater freedom of choice of the lessors as well as the ability of many small equipment firms to develop certain items at least as effectively as the Bell System. It reflects too the Bell System's private interest in minimizing product proliferation and in treating its equipment and optional services as the more lucrative side of its business. The changes since 1965 have unlocked innovative forces which the consent decree had helped to restrain; the gains in innovation may prove to be substantial.[52]

Some Bell innovations have probably been diffused more slowly because of the decree and the long-standing habits that it hardened. For example, Bell System work on picturephones is said to have brought advances in television technology, but use of these advances in nontelephone markets is restrained, at least for the time being, by the inability of Western Electric to enter them in general market competition. The possibilities for commercial application of Bell System capabilities in computer design and development have almost certainly been significantly less than they would have been without the consent decree. Little has been forthcoming since the Bell computer prototypes of the 1930s and high-speed "Leprechaun" of 1957. Such examples do not permit quantitative estimates of the net innovation loss but indicate that it has not been trivial. And the loss is more likely to grow than to shrink as the scale and variety of communications apparatus enlarge further.

Exclusivity of Technology

As was noted under the third hypothesis, one obvious and important facet of the interaction of innovation and market structure is a prefer-

52. See also the development by the General Electric Company of TermiNet 300, an advanced challenger to the Bell System's teletypewriter terminals (made by the Teletype Corporation, a subsidiary of Western Electric), which have held nearly all of the market (*Business Week,* April 5, 1969, p. 52); Litton Industries' development of an interconnectable in-flight telephone (*Wall Street Journal,* May 6, 1969); and IBM's new offering of an automatic telephone and data switching exchange (*Business Week,* April 19, 1969, p. 39).

ence for ownership of capacity. Besides adding rate base, ownership gives the carrier control over access to the system and the direction of technology pertaining to future exclusivity. The ownership preference affects a range of inventive and innovative choices made by the integrated carriers; the resulting innovations and the market structure are consequently more exclusive than is necessary.

The carriers' treatment of terminal attachments probably in part reflects a preference for exclusivity. The Bell System averted all nonacoustical foreign (non-Bell) attachments until 1965 by mere insistence, without any quantitative analysis or any evidence of the net variance from optimum (or merely acceptable) service standards that they might cause. Only after an unexpected FCC order in 1968[53] posed that issue explicitly did the Bell System reveal that it had ready an inexpensive interfacing device that, as far as technology was concerned, could easily be applied to make possible the removal of virtually all prohibitions on attachments. One can reasonably infer that for years, possibly decades, the System has exaggerated the technological case against foreign attachments and has neglected research and development on devices and system designs that would accommodate such equipment.

Perhaps the most obvious indication of a preference for exclusivity is the vertical integration in the Bell System. This issue has several facets, both institutional and technological. Whatever its benefits, integration prevents any direct objective comparisons or tests of the Bell System's R&D and production capabilities. No serious observer has denied that the "teamwork" arising from integration provides genuine advantages, but the System has yet to offer any quantitative evidence about these benefits that could be compared with the possible costs.[54] Merely to list innovations promoted by teamwork is of little or no value in identifying the net effects of integration.[55] Could the innovations have been more frequent, better, and more varied? The Bell System has not discussed in practical terms the various alternatives—such as partial integration, long-term contracting, and joint ventures, varying according to the equipment types and subject to experimentation and change—that are available, many of

53. See note 34.
54. The recent McKinsey *Study of Western Electric's Performance* provides very little hard evidence.
55. "Vertical Integration in the Bell System" contains some qualitative statements about the benefits and listings of innovations, but without a normative basis for appraising them.

which are fully operative in a host of other sectors and countries. Instead it chooses to frame the issue as a choice between all vertical integration and none, between "reality and theory." In fact, almost total integration is unusual among all utility and industrial sectors in the United States and abroad, even in sectors where technology is at least as complex as it is in communications and where reliability is at a premium, as it is in aircraft and nuclear power. It is reasonable to rest the burden of proof on Bell's extreme form of exclusivity. International comparisons may be helpful, but they must be approached with great care.[56]

As a final point, integration has not in recent years been valued by the Bell System primarily because it serves a profit-maximizing strategy to escape regulatory control. A frequently expressed fear is that the Bell System would set high equipment prices as a way of realizing high total profits and escaping regulatory constraint. Yet rather than reap excess profits through Western Electric, whose reported after-tax rate of return on assets has recently been between 9 and 10 percent, Bell has probably set equipment prices marginally on the *low* side. The issue of vertical integration, therefore, pertains essentially to a straightforward trade-off between technical gains and competitive losses. The Bell System's tenacious defense, so devoid of data about these trade-offs, creates the suspicion that the principal usefulness of integration is in reinforcing exclusivity.

The point is not so much that the losses actually outweigh the gains— that is now impossible to demonstrate by an independent appraisal. The point is rather that the carrier has avoided subjecting the trade-off question to adequate review and has provided virtually no quantitative evidence about the alternatives. This behavior is consistent with the

56. The British telephone system's lackluster results from buying from a group of outside suppliers is probably of little relevance to the American situation. A more aggressive purchasing approach by the General Post Office, which operates the telephone system, could have improved the results substantially, and if vertical integration *is* beneficial, it may be enough for the Post Office to acquire only one of the suppliers; see my chapter in Caves and Associates, *Britain's Economic Prospects.* One curiously unreliable source on the question is the otherwise excellent report, *The Post Office,* by the Select Committee on Nationalised Industries, House of Commons 340 and 340-1 (London: Her Majesty's Stationery Office, 1967). Several committee members made a brief trip to the United States but visited only the Bell System. Not surprisingly, the report includes some brief references to the advantages from integration said to be enjoyed by the Bell System. But these remarks were not based on any evaluation or even a brief exposure to other points of view on the issue, even though they are prominently quoted twice in "Vertical Integration in the Bell System."

motivation of maximizing the technical and institutional exclusivity of the System.

Among the most important problems in exclusivity during the next ten years will be the domestic satellite system. Before the end of the 1970s, large satellites with refined directionality can be developed, capable of at least partially supplanting other modes of broadcasting and point-to-point transmission. In effect, an alternative (and also complementary) system to both the Bell cable and microwave system and the present network structure of television broadcasting (and possibly community antenna television) is near at hand. One can envision a wholly independent satellite entity, offering circuits to all comers, including jobbers, the Bell System, broadcasters, and large and small enterprises; or possibly a multiplicity of satellite entities with diverse capabilities, growing out of proposals since 1965 by the American Broadcasting Company, the Ford Foundation, and others.[57]

The range of technical and ownership possibilities is very wide, and it offers an important chance to gain the benefits of competition in communications without damaging "utility" functions of established carriers. In early 1970 the White House recommended leaving the domestic satellite field open for entry. If the FCC follows this course, the result will be a distinct shift away from systemic exclusivity, as long as entry is genuinely open and actually tried by many firms.

Preemptive Innovation in Response to Competition

If there is a threat of new entry, through innovation, into a firm's traditional markets or into attractive new ones, the firm will increase its R&D activity and innovation in those directions in an effort to prevent entry—possibly to the point of negative social returns. On the other hand, a firm's known ability to preempt the field may forestall any such R&D activity by others, leading to little or no innovation.

While no general evaluation of each of these effects in the communications field is possible, the experience with microwave innovation is enough to suggest that they can be powerful.

Microwave technology dates back to well before its first public demonstration in 1915, and a variety of public and private groups—American and British—participated in its development during the 1930s and

57. See Lessing, "Cinderella in the Sky," p. 132.

World War II. By 1944, microwave had emerged as the obvious technique for meeting the massive increase in broad-band intercity transmission capacity that television would require in the postwar years. By 1946, several firms had created or planned microwave capacity between major eastern cities and had applied for an FCC franchise, with further expansion in prospect.[58] The Bell System recognized the advantages it could gain in displacing these independent systems, and it proceeded to do so with utmost thoroughness. It mounted a large, rapid two-stage innovation effort between 1946 and 1950 to create a nationwide microwave relay system—called the TD-2—to preempt the microwave field. This involved a crash program to make the technology operational quickly, and the new system displaced much of a large expansion of coaxial cable capacity that had been projected by the Bell System in 1946. The program was extensive and costly, and some specific expenditures no doubt yielded a low social return, even though they evidently generated high private returns to the Bell System by helping to assure exclusivity in this crucial area. The development was telescoped into about three years, and the Bell System had acted quickly enough to take over virtually the entire domestic microwave carrier field. Meanwhile, AT&T unremittingly refused to interconnect with rival microwave systems. As on other issues, the refusals were not supported by quantitative evidence about quality of service and other marginal costs and benefits that might be involved. By the period 1949–50 the Bell System had succeeded in maneuvering the FCC into converting the microwave field—with much potential for fruitful competition—into a total monopoly, not only of transmission operations but also of equipment manufacturing. Indeed, control extended even further. The FCC excluded all private microwave until 1959. Only when the FCC's "Above 890" decision removed that exclusion did the Bell System respond with the TELPAK offering and related data-transmission services (see the next hypothesis).[59]

After 1952 microwave innovation reverted back from costly and uncharacteristic "revolution" to the System's traditional approach of

58. See the study, "The Development of the TD-X and TD-2 Microwave Radio Relay Systems in Bell Telephone Laboratories," Weapons Acquisition Research Project, Harvard Business School (October 1960; mimeographed); A. C. Dickieson, "The TD2 Story: Changing for the Future," *Bell Laboratories Record*, Vol. 45 (December 1967), pp. 357–63; and Donald C. Beelar, "Cables in the Sky and the Struggle for Their Control," *Federal Communications Bar Journal*, Vol. 21 (January 1967), pp. 26–41.

59. See Phillips, *Economics of Regulation*, p. 677.

"evolution." The subsequent improvements have involved frequent cost overruns and slippages of schedule, in contrast to the extreme speed and precision with which Bell established TD-2 under competitive pressure.[60] Yet for low-density-route, short-haul radio systems, which are under competitive pressure, the Bell System has again responded quickly, with tight scheduling and generous R&D resources. During the 1960s, renewed efforts to open trunk-route microwave transmission have been deflected by the FCC's reluctance to order the Bell System to interconnect with independent systems. And Bell has imitated consistently—and filed exclusive tariffs for—new service offerings by potential entrants wherever they have begun to attract the serious interest of the FCC.[61]

The whole microwave issue may soon fade if satellites, improved cables, and other techniques reduce the role of microwave. As a result, the Bell System may be forced to yield its grip on the microwave field only after microwave ceases to have a significant effect on its rate of return or risk.

The shift from coaxial cable to microwave after 1946 might be regarded as belying rate-base preference (third hypothesis), because microwave was less capital-intensive than coaxial technology. On the other hand, the shift may simply underline how powerful the preemptive innovative response was—enough to override any rate-base preference.

One example alone, even so graphic and repetitive a one as microwave, cannot pretend to establish a general pattern of preemptive innovation. But it does indicate that regulatory procedures may be used adroitly by the carrier to control market structure. Currently, domestic satellites offer similar possibilities for increasing the scope of carrier monopoly, although a full understanding of the alternatives has not been attained. Even if sufficiently alert to this situation, the FCC may fail to explore or to clarify the whole range of technological possibilities and thus limit innovation and promote monopoly.

60. Weapons Acquisition Research Project, "Development of the TD-X," Chap. 6; and "Development of TH and TJ Microwave Radio Relay Transmission Systems." The TH system "slipped approximately five years from the original completion date of December, 1955, established in August, 1952." The lack of competitive pressures for TH appears to have been partly responsible for the difference. See also Marschak in *Rate and Direction of Inventive Activity.*

61. For example, a proposal by Microwave Communications, Inc. (MCI), to develop a network for educational broadcasting prompted the Bell System in 1969 to cut its rates for such service sharply (FCC Docket 18316). See letter from John D. Gocken of MCI to Rosel H. Hyde of the FCC (April 22, 1969) detailing the Bell System's procedural tactics in the matter.

Entry into Unprofitable Markets

Limitations on rates of return may induce a utility to enter additional markets with submarginal returns, if the necessary capital can be included in the rate base and if the allowed rate of return exceeds the cost of the capital.[62] Even without regulation a monopoly is sometimes encouraged to incur losses in one market in order to facilitate its own entry into other markets as part of a general price-discriminating strategy. Regulation adds incentives for such behavior, possibly even promoting some entry with negative returns, unless it explicitly screens individual rates of return and rate-base inclusion for such activities. Two probable examples are examined here: private-line services, and TELPAK and related business services. Hearings by the FCC amassed evidence on these, though no immediate action was taken.

During the years 1956–63, the FCC investigated private-line services (telephone, teletypewriter, telephotograph, remote metering, and other signaling services) in great depth. It found that subnormal rates of return (3 percent) were being earned on services under competitive pressure, balanced by higher rates (11 percent) on protected services.[63] The commission's orders for rate revisions to equalize returns were appealed in federal court by some previously favored users, but the orders were upheld.[64]

The Bell System responded to new private microwave competition after 1959 with several new business services, primarily TELPAK, WATS (wide area telephone service), and TWX (teletypewriter exchange service). Competitors challenged these services, alleging that they were priced below long-run marginal costs. Studies for the FCC did suggest that some of them, especially TELPAK, were drawing low rates of return. The Bell System calculated and, in compliance with an FCC request, recalculated the returns on the services, presumably using accounting and cost allocations to minimize the differences in rates of return as much as possible. Apparently TELPAK earns a very low rate of return, possibly less than 1 percent.[65] Not surprisingly, TELPAK has attracted

62. Averch and Johnson, "Behavior of the Firm"; and Posner, "Natural Monopoly."
63. FCC Dockets 11645 and 11646, Jan. 30, 1963.
64. *Wilson and Co., Inc.,* v. *United States,* 335 F 2d 788 (1964).
65. See FCC Docket 14251, and Docket 14650, Exhibit AT&T 82, which describes the so-called Seven-way Cost Study prepared by AT&T. This study analyzed costs of seven major service categories.

hundreds of major customers, and these form a powerful opposition to increases in TELPAK rates. After five years of efforts by the FCC to get adequate revisions by the Bell System, TELPAK rates are still the subject of sharp controversy.[66]

Other examples of low-return services include service in small towns and certain peak-load services. Not all are evidence of excessive entry; only those that provide entry into relatively new and lucrative markets confirm the seventh hypothesis. Recently the FCC has required the Bell System to provide evidence of returns on a wider range of services. Even before these data have been evaluated, there are strong indications that the hypothesis has been valid in recent years on a significant scale.

Implications of the Seven Hypotheses

The indications above may be only exploratory and interpretive, but they are also consistent and not trivial. The structural interactions appear to be clearer and more important than the behavioral effects, except possibly for forestalled innovation. From one point of view, one could say that FCC actions affecting structure have probably affected innovation and performance a good deal more than have the rate-of-return constraints that the FCC and the state commissions have tried to apply. From another point of view, one might say that the Bell System in particular has had more to gain by reducing risk through strategic moves (including "portfolio" selection among innovations) than by trying to get an increase in the permitted rate of return.

The seven hypotheses suggest divergences from the optimum pattern of innovation, but some of the effects may, in practice, not create divergences. The carriers' temptations to engage in restrictive pricing and investment appear to have been generally overcome (at least there is no strong contrary evidence), and the effects of rate-base preference may actually have hastened at least some capital-embodied innovations.[67] Yet not all the behavioral effects fit these positive welfare interpretations, and the total net effects will not be known without further study.

Over all, the structural effects have probably included (1) a distinct checkering of carrier innovations, with relatively slow and modest "evo-

66. *Wall Street Journal*, April 28, 1969.
67. See especially Alfred E. Kahn, "The Graduated Fair Return: Comment," *American Economic Review*, Vol. 58 (March 1968), pp. 170–74, and "Inducements to Superior Performance: Price," in Harry M. Trebing (ed.), *Performance under Regulation* (Michigan State University, Institute of Public Utilities, 1968).

lutionary" changes punctuated by specific crash programs; (2) technology that is on the whole more carrier-exclusive than is either optimal or necessary; (3) a significant amount and variety of forestalled innovation by independent firms who see no chance of access; and (4) generally greater carrier control of market structure through adroit use of regulatory policy than is consistent with efficiency and innovation. The ultimate net avoidable losses from these effects may not be massive, though they are likely to have been significant. One *can* say that quite a few marginal changes in the structural features of regulation could now yield important gains and that some substantial changes might yield surprisingly large net gains.

The partitioning off of Western Electric under the consent decree invites further study. The gains from coordination would have to be large to outweigh the possible net losses, so information about these magnitudes would be valuable. The Justice Department was said to be reviewing the question in 1968, but a larger research effort will ultimately be needed. Although the Task Force on Communications Policy effectively prevented any action on the issue during its August 1967 to December 1968 period of work, it added little new evidence for independent appraisal, and so the time interval was lost. The situation is evidently stable; neither side is likely to press for change, and even the more powerful equipment manufacturers are avoiding direct challenges through major equipment offerings.[68]

Some recent FCC moves, such as those on Carterfone and MCI, and the White House proposal for competition in domestic satellites suggest a growing awareness that structural regulation offers a powerful means of control and an increased disposition to apply it. But the practical scope for changes by the FCC may be small, even if it behaves as a well-informed and independent agency, as long as there is no powerful impetus for change and no independent source of thorough quantitative information.

68. An example of this is IBM's new automatic telephone and data switching exchange for businesses, the 2750, which is being offered quietly and solely in European markets (*Business Week,* April 19, 1969, p. 39). IBM continues to avoid any attempt to enter what could be a large new market for its equipment. And while General Electric has now challenged the Teletype Corporation, it has not attempted to capture the really substantial equipment markets of Western Electric.

Policy Implications

It is perhaps too facile to suggest that the whole range of opportunities to introduce competition should be evenly cultivated but without threatening the carriers' existence. Nonetheless that *is* a proper ultimate goal for regulators, in order to counterbalance the natural tendency for regulation to be manipulated—in some cases through biases in innovation—into permitting more market power than is necessary.

From this point of view, the basic strategy of the FCC may have been defective and self-defeating. The policy of continuous surveillance appears to have caused at least some of the predictable economic side effects; also it has surrendered the leverage provided by open hearings and independent study. Furthermore, it has made a thorough, objective appraisal of the FCC's effects impossible.

A plausible conclusion is that, on balance, the FCC has shielded the carriers structurally without effectively controlling or even clarifying their behavior. This has led to weak profit and price constraints, economic side effects, and a number of questionable structural decisions (which, in turn, have further undermined the footing for regulation). There is wide agreement that a more effective approach would have opened more competitive possibilities, thereby reducing the need for behavioral constraints. And in fact the FCC has moved in this direction since the middle 1950s. Paradoxically, this shift in emphasis may foster just that withering of explicit regulatory constraints that is appropriate in the long run.

State regulation has been rather more passive. Some aspects of it appear to be ripe for change, but there is little impetus for revisions at present. State commissions could at least supplement FCC structural moves rather than resist them, as they frequently have done in the past.

A stiffening of regulation tends to stir pressures for intervention from Congress and the executive branch where carrier and related interests are relatively strongly represented. The Defense Department has played a key role in some structural questions. Proposals based on national defense "needs" have rarely been costed out and weighed against all alternatives; often they have simply prevailed *in camera* without objective appraisal.

At present, several specific policy moves are needed to improve innovation as well as other aspects of performance. Most of them concern structure, and some of them go beyond what has come to be regarded as the customary jurisdiction of the FCC and the state commissions.

One possible step is a full review of the merits of removing all or part of the consent-decree partition between the Bell System (including Western Electric) and the rest of the communications equipment market. The standard antitrust proposal is to remove Western Electric from Bell and separate it into two, or possibly three, independent firms. This may be technologically feasible, since Western Electric has two or more plants making most major components, and others could be added. The costs of transition would probably be substantial, but they might be small in comparison with the total long-run gains. By using imaginative and aggressive purchasing policies, the Bell System could eventually develop alternative domestic and foreign sources for most or all of its major equipment components. The FCC and state commissions might need to monitor contracting activity for some time, but the change would probably soon yield a crop of substantial competitors for many, if not most, significant contracts.

Actually many changes in Western Electric's status other than just the traditional remedy are possible, even though these choices have not been fully considered. In any event, if the Gordian knot represented by the decree is ever to be cut, a Justice Department suit is probably the best vehicle for giving the issue full review. Trial and remedy would take time, perhaps five to ten years, but enough appears to be at stake to warrant the effort, and the stability of the status quo suggests that, unlike many other antitrust problems, this one is not self-correcting. Apart from providing benefits in innovation and efficiency, the change would enlarge regulatory information while reducing the need for many of the present regulatory controls.

Whether or not the consent decree is to be expunged or changed, a number of other regulatory steps would help to foster moderate and consistent competitive pressures. The most important is a deliberate attempt to maximize the independence of domestic satellites from Bell System control, both in development and in management, and to have as many separate participants as possible; this is in fact the direction of recent policy. A range of alternatives is available, among which Comsat appears to provide nearly the *least* possible independence and diversity of participation. The FCC may not be capable of taking the initiative to create the best feasible structure; so far it has taken the passive stance of dealing only with proposals set before it. Indeed, one major policy lesson is that competitive possibilities have been routinely lost by default because the FCC's powers, resources, and perspective are too narrow for many of the tasks thrust upon it.

The FCC could require substantial experimentation, at least, in independent microwave systems for both commercial and public broadcasting users.[69] Under proper safeguards for interconnection, this area could be enlarged considerably without damaging any essential Bell System interests. The FCC has imaginatively anticipated the existing inducements for turning computer time-sharing into a common carrier, and hence a monopoly, service.[70] But these inducements are not disappearing, and the FCC will need to watch the problem closely.

The Bell System's custom of not opening equipment contracts to outside competition apparently has no basis in the consent decree or in other law. Certainly some parts of the Bell System equipment market could be opened at least experimentally for competitive bids. The results would be mixed and perhaps negative at first, since (1) the Bell System would have incentives, as rational maximizers, to select those equipment items on which its present cost advantages are great; (2) loss-leader pricing would be alleged by both sides; and (3) independents might be reluctant to challenge on any substantial items, for fear that doing so would eventually lead to abandonment of the whole decree. Even so, the experiment might open up some potential competition that is now wholly excluded. In addition, methods for diffusing Bell innovations more rapidly outside the System could be explored more fully.

Other specific steps may be taken over the years, especially if the FCC is willing to investigate other barriers to entry that have no clear technological basis. Unfortunately, the state commissions are nearly powerless to effect most of the significant structural changes that would be desirable. Even so, they might assist in challenging some barriers (or at least not defend them); for example, a state commission might require an op-

69. See particularly the MCI case, FCC Dockets 16509–16519 (referred to in note 14), which has been decided, as a matter of principle, in MCI's favor. The ultimate effect of this decision remains to be seen.
70. By inviting all interests to express views in 1968, the FCC brought out the strong case for not putting the service under carrier control. See especially the Justice Department submission, FCC Docket 16979 (cited in note 14).

erating carrier to open up some equipment contracts to outside competition.

Even apart from policy changes, further research is needed on a variety of trade-offs among technological factors, such as system integrity, reliability standards, and economies of vertical and horizontal integration. Costs and rates of return on specific services need further study, and virtually no independent analysis at all has been made of the comparative costs of equipment under hypothetical open-supply arrangements. The role of federal R&D support and its various spillovers also needs clarifying, and much may be learned from analyzing the varying innovativeness of the Bell operating subsidiaries. These and other research questions are interesting and important; where there is no competition, they possess a degree of urgency.

Evidently the alternatives that technology offers in communications have not been clearly understood and analyzed under past regulation. In turn, the possible relations between technological change and both the practice and the lacunae of actual regulation are also conjectural, even though some of the probabilities can be indicated. The economist's familiar concluding pleas for more and better data and for more research— much more research—are in order.

CHAPTER FIVE

Air Transportation in the United States

Almarin Phillips

COMMERCIAL PASSENGER air transportation began in the United States after, and as an immediate consequence of, the Air Mail Act of 1925, known as the Kelly Act, and the Air Commerce Act of 1926, although some generally unsuccessful attempts at entry by a small number of carriers had been made a few years earlier. The Kelly Act provided that contracts for carrying airmail be awarded to private companies; the Air Commerce Act required the federal government to develop and maintain a system of airways and navigation aids and to certificate aircraft and airmen in the interests of safety. Thus, from the very beginning, the airline industry has been affected by some form of federal intervention, if not by direct regulation.

In the forty-odd years of its existence, the industry has experienced more continual, important technological change than any other regulated industry. The beginning of regulation and the periodic changes in its nature are related to the technological changes in the industry. But this study argues that, in general, regulation in the industry has adjusted (though with substantial time lags) to the circumstances created by changes in technology. While regulation has had some effect on certain innovations in the commercial aviation industry, the rapid pace of the more fundamental innovations has been made possible by technological developments outside the industry, for reasons usually unrelated to commercial air transportation. Regulation has had little effect on technology up to now, though this may not be the case in the future. Coming changes in commercial aviation technology seem likely to depend peculiarly on the research and development (R&D) policies of transportation regulatory agencies.

The technological changes that have occurred in commercial aviation are described below in some detail, and their importance for the growth of the industry is discussed. Until recent years, technological changes have tended to increase the minimum efficient size of air carriers nearly as rapidly as the market was expanding. This affords a view of the lag in public regulation, the early efforts at private instead of public regulation, and the pervasiveness of the interests of the principal carriers in public regulatory policies.

Evidence is also offered on the limited role the carriers and regulatory agencies have played in the technological changes in the industry. The primary sources of significant technological advances are indicated, and airline R&D programs are described and related to these advances. The effects and purposes of these programs are compared with the effects and purposes of R&D from other sources. Finally, changes that are taking place in aviation research and development are noted, and questions are raised as to whether more explicit attention needs to be paid to R&D for air transportation in the immediate future.

Extent and Market Consequences of Technological Change

Technological change in aviation can be viewed in a number of ways. One examines changes in the performance characteristics of the aircraft used by commercial carriers over the years. These characteristics can be sketched fairly clearly for the years since 1932. For the years before 1932, information on individual aircraft types and their performance is available, but data on the number used by the different airlines are neither complete nor reliable.[1]

1. I have had to use my own judgment in deciding which data are the most nearly accurate, even for the 1930s. As the *Handbook of Airline Statistics* states concerning aircraft: "Literally dozens of sources have been consulted, but multiplicity of sources often produced a multiplicity of answers to the same question. Part of the difficulty lies in the fact that any given plane model could and did vary as customer specifications varied. Moreover, models were often modified after delivery. . . . Thus, for a given model, engines can vary as to manufacture and power, and power ratings seem to be in constant flux; number of seats can differ astonishingly; 'normal cruising speed' is a most ticklish item to nail down; 'range' is even more difficult; 'cargo capacity' is often just not obtainable; 'price', understandably, varies with quantities purchased, and even more with date of purchase; and 'date of first introduction into scheduled airline service' makes one wonder if accurate history can ever be written" (Civil Aeronautics Board, *Handbook of Airline Statistics* [1966], p. 510). Much the same situation exists for the airlines.

Characteristics of Aircraft

In late 1932, 456 fixed-wing aircraft were in service or available for service by scheduled domestic airlines.[2] The list below shows that 374 of these were land-type aircraft, with a seating capacity of four or more; they were owned by seventeen of the twenty-nine lines that were in operation between July and December of that year.[3] The remaining 82 aircraft were smaller planes or flying boats owned by the same lines or were unidentifiable other types owned by the twelve other carriers.

Manufacturer and model	Number in fleets
Ford (Stout) Tri-Motor	63
Boeing 40C and 40-B4 (single engine)	57
Stinson Detroiter and other single engine	32
Fokker F-10 (trimotor)	28
Stinson SM-6000 and U (trimotor)	26
Fairchild 71 (single engine)	20
Fairchild Pilgrim 100 (single engine)	20
Fokker Super Universal and other single engine	19
Lockheed Vega (single engine)	18
Boeing 80 and 80A (trimotor)	12
Lockheed Orion (single engine)	12
Travel Air (single engine)	12
Curtiss-Robertson Kingbird (two engines)	9
Fokker F-32 (four engines)	9
Hamilton Silver Streak (single engine)	9
Metal Aircraft Flamingo (single engine)	9
Northrop Alpha (single engine)	6
Curtiss Condor Transport (two engines)	5
Consolidated Fleetster (single engine)	3
Bellanca Pacemaker (single engine)	2
Boeing 200 and 221 (single engine)	2
Ryan Brougham (single engine)	1
Total	374

Source: *Aero Digest* (July, October, November 1932); *Aviation*, Vol. 31 (August 1932), pp. 344–47; *Moody's Manual* (New York: Moody's Investors Service, 1933).

2. *Aviation Facts and Figures,* 1957 ed. (Washington: American Aviation Publications, 1957), p. 90.

3. The seventeen lines were American, Bowen, Braniff, Eastern, Hanford, Ludington, National Parks, Northwest, Pennsylvania, Rapid Air, Transamerican, Transcontinental and Western Air, United, United States, Varney, Western, and Wyoming. See *Air Commerce Bulletin* (March 1, 1933), pp. 421–23, and sources given for list.

Table 5-1. Technical and Performance Characteristics of Aircraft in Domestic Fleets, 1932

Manufacturer and model	Year entered service[a]	Engines		Typical seating capacity	Gross takeoff weight (pounds)	Typical cruise speed (miles per hour)	Normal full-load range (miles)	Wing loading (pounds per square foot)
		Number	Aircraft horsepower					
Bellanca Pacemaker	1929	1	300	5	4,300	110	900	15.7
Boeing 40-B4	1929	1	525	4	6,075	110	535	11.2
Boeing 80A	1929	3	1,575	16	17,500	125	460	14.0
Boeing 221	1930	1	575	6	8,000	137	500	15.0
Consolidated Fleetster	1930	1	575	7	5,600	150	500	17.9
Curtiss Condor Transport	1929	2	1,200	19	17,380	116	510	11.5
Curtiss-Robertson Kingbird D-1	1930	2	480	6	6,115	120	450	15.1
Fairchild 71	1928	1	425	6	5,500	110	650	16.6
Fairchild 100	1930	1	575	8	6,500	122	450	15.6
Fokker Super Universal	1928	1	425	6	4,750	118	735	15.0
Fokker F-10	1928	3	1,275	10	13,100	118	765	15.4
Fokker F-32	1930	4	2,100	30	22,500	120	—	16.7
Ford Tri-Motor 5-AT-C	1929	3	1,350	15	13,500	113	525	16.1
Hamilton Silver Streak	1928	1	400	7	5,750	115	600	14.3
Lockheed Vega (5C)	1930	1	420	6	4,500	150	700	16.4
Lockheed Orion	1931	1	450	6	5,200	180	600	18.7
Metal Aircraft Flamingo	1929	1	510	7	6,000	115	580	16.8
Northrop Alpha	1930	1	420	6	4,500	140	700	15.3
Ryan Brougham	1929	1	300	5	4,000	110	800	14.3
Stinson Detroiter SM-10	1928	1	220	5	4,500	115	700	14.7
Stinson SM-6000	1930	3	645	9	8,500	—	—	17.3
Stinson U	1932	3	720	9	9,300	119	400	16.2
Travel Air A-6000-A	1929	1	420	6	5,500	120	680	16.2

Sources: *Automotive Industries*, Statistical Issues (Philadelphia: Chilton Class Journal Co., 1928–32); *Aviation* (January 1930, 1932, 1933); *Air Commerce Bulletin*, various issues.

a. The year shown is for the model indicated. The first model of the Lockheed Vega was used in 1928; of the Stinson Detroiter in 1926; of the Fokker Universal in 1926;

The technical and performance characteristics of the 374 aircraft—which included the largest and most modern planes available at that time—are shown in Table 5-1. Specifications for the early aircraft are less definitive than at first appears. Horsepower is usually, though not always, reported as maximum takeoff power. Cruise speed appears often to be given as the maximum technically sustainable speed rather than as a slower, more typical and more economical cruise speed. Range data seem sometimes to approximate the full-fuel range rather than the full-passenger-load range.

In brief, 222, or nearly 60 percent, of the planes were single engine. The plane in greatest use was the Ford Tri-Motor, the most popular model of which had seating for fifteen passengers. The Ford had a cruise speed of about 113 miles an hour and a normal range of 525 miles. The fastest plane in use was the Lockheed Orion; it also had a wing loading of 18.7 pounds a square foot, the highest of the group. The largest was the Fokker F-32, with a gross weight of over 11 tons and seats for thirty passengers.

Thirty-three years later, at the end of 1965, the fleets of the eleven domestic trunk-line carriers totaled 1,122 fixed-wing aircraft. The list below gives the breakdown by type of plane, and Table 5-2 shows their

Manufacturer and model	Number in fleets[a]
Boeing 727	167
Douglas DC-6, -6A, and -6B	147
Boeing 707	125
Lockheed L-188	117
Boeing 720 and 720B	114
Douglas DC-8	99
Convair 340 and 440	67
Vickers Viscount V-745 and V-812	55
Convair 880	46
Lockheed L-1049 (B through G)	46
Douglas DC-7, -7B, and -7C	44
Lockheed L-049, L-649, and L-749	29
Sud Aviation Caravelle SE-210	20
Convair 990	18
British Aircraft Corporation BAC-111	12
Douglas DC-3	7
Lockheed L-1649	5
Douglas DC-9	4
Total	1,122

Source: Federal Aviation Agency, *FAA Statistical Handbook of Aviation, 1966* (1967), pp. 139–45.
a. Includes cargo versions of passenger-type aircraft.

Table 5-2. Technical and Performance Characteristics of Aircraft in Domestic Fleets, 1965

Manufacturer and model	Year entered service[a]	Engines Number	Engines Total power	Typical seating capacity[b]	Gross takeoff weight (pounds)	Typical cruise speed (miles per hour)	Normal full-load range (miles)	Wing loading (pounds per square foot)
Boeing 707-120	1958	4	54,000[e]	132	285,000	550	3,100	105.6
Boeing 720B	1961	4	72,000[e]	120	235,000	560	4,100	96.2
Boeing 727-100	1964	3	42,000[e]	110	160,000	570	2,100	97.0
British Aircraft BAC-111-200	1965	2	20,800[d]	79	78,500	550	1,500	78.2
Convair 340	1952	2	4,800[d]	44	47,000	280	700	51.1
Convair 880	1960	4	42,400[e]	110	184,500	556	3,500	92.5
Convair 990-30A-6	1963	4	64,000[e]	110	255,000	605	4,700	113.3
Douglas DC-3	1936	2	2,400[d]	21	25,200	175	850	25.5
Douglas DC-6B	1951	4	10,000[d]	66	107,000	300	2,800	73.1
Douglas DC-7-120	1953	4	13,000[d]	76	122,000	350	3,500	83.5
Douglas DC-8-10	1959	4	54,000[e]	132	273,000	544	4,300	100.2
Douglas DC-9-10	1965	2	24,000[e]	90	77,700	560	1,500	83.2
Lockheed L-749	1947	4	10,000[d]	55	107,000	285	2,600	64.8
Lockheed L-1049B	1951	4	11,200[d]	83	120,000	300	2,900	72.7
Lockheed L-1049G	1955	4	13,600[d]	80	137,500	310	4,100	83.3
Lockheed L-1649	1957	4	13,600[d]	75	160,000	310	4,900	86.5
Lockheed L-188	1959	4	15,000[d]	85	116,000	405	1,600	89.2
Sud Caravelle SE-210	1961	2	25,200[c]	64	110,230	505	2,300	69.8
Vickers Viscount V-745	1955	4	6,680[d]	48	64,500	315	1,500	70.0

Sources: *Aviation*, various issues; *Aviation Week and Space Technology*, Vol. 82 (March 15, 1965), p. 196; *Flight*, Vol. 76 (Nov. 20, 1959), pp. 597, 606, 611; *Flight International*, Vol. 88 (Nov. 25, 1965), pp. 926–29; *1966 Aerospace Year Book* (Aerospace Industries Association of America, 1966), p. 286.
a. Year specific model entered service in U.S. domestic airlines.
b. Seating capacity is that typical in year of introduction, with medium-density, mixed-class service for aircraft introduced after 1953. All-coach, high-density seating may increase seating capacity substantially beyond these figures. The DC-6B, for example, may accommodate up to 107 passengers in the highest-density version.
c. In pounds of thrust for turbojet and turbofan engines.
d. In horsepower for piston and turboprop engines.

operating characteristics. All were multiengine aircraft. About half were powered by turbojet or turbofan engines. The DC-8, the heaviest type in domestic service in 1965, had a gross takeoff weight more than ten times that of the Fokker F-32. (International versions of both the DC-8 and the Boeing 707 are larger and heavier than those shown in Table 5-2.) The DC-8 could carry more than four times as many people nearly five times as far per hour of flying as the F-32. Each square foot of wing area of the DC-8 or of the Boeing 707 lifted at takeoff about six times the weight lifted by a square foot of wing area of the F-32. The aircraft of 1965 that were in long-range, transcontinental use had ranges more than five times those of the aircraft in use in 1932. Indeed, the short-range and intermediate-range planes of 1965 had ranges greater than those of the long-range planes of 1932.

Overall Qualitative Changes in Air Travel

Another way of viewing the extent of technical changes is by overall effects on the nature of air travel. These can be gleaned from Table 5-3. On an average, an air traveler in 1965 rode in a plane that had almost fourteen times as many seats, flew three times as fast, and had perhaps one-fiftieth[4] the chance of a fatal accident per mile traveled as its counterpart in 1932. Scheduled airlines could fly nearly 105,000 route-miles in 1965, in contrast with fewer than 29,000 miles in 1932. In other words, technology was providing a product in 1965 that was very different from that of three decades earlier.

Operating Costs

Descriptions of changes in the technical and qualitative characteristics of aircraft and air travel do not show the effects of technological change on the costs of providing air transportation. Estimates of operating costs per seat-mile for various aircraft in use since the late 1920s are given in Table 5-4.[5] Operating costs include expenses of flying operations (fuel, oil, flight crew wages and salaries, and so forth), and maintenance and

4. Based on the mean fatality rate per passenger-mile for 1932–34 compared with that for 1963–65. Because most accidents occur during the takeoff and landing phases of flights, that may not be the best basis for comparison.

5. The data in Table 5-4 are adapted from the author's *Technological Change and Market Structure: A Study of the Market for Commercial Aircraft* (Heath, 1970). A sample of cost observations was taken, with permission, from Ronald E.

Table 5-3. Changes in Characteristics of Air Travel on Scheduled Domestic Air Carriers, 1932–65

Year	Average seating capacity	Average airspeed[a] (miles per hour)	Total route-miles	Passenger fatalities per 100 million passenger-miles
1932	6.6	109	28,956	14.9
1933	7.6	116	28,283	4.6
1934	8.9	127	28,609	9.0
1935	10.3	142	29,190	4.7
1936	10.7	149	29,797	10.0
1937	12.5	153	32,006	8.3
1938	13.9	153	34,879	4.5
1939	14.7	153	36,654	1.2
1940	16.5	155	42,757	3.0
1941	17.5	160	45,163	2.3
1942	17.9	159	41,596	3.7
1943	18.4	154	42,537	1.3
1944	19.1	156	47,384	2.2
1945	19.7	153	48,516	2.2
1946	25.3	169	53,981	1.2
1947	29.9	170	62,215	3.2
1948	32.4	176	68,702	1.3
1949	34.7	178	72,667	1.3
1950	37.1	180	77,440	1.1
1951	39.1	183	78,913	1.3
1952	42.2	189	77,894	0.4
1953	45.6	196	78,384	0.6
1954	49.6	204	78,294	0.1
1955	51.5	208	78,992	0.8
1956	52.1	210	84,189	0.6
1957	53.7	214	87,550	0.1
1958	55.5	219	89,569	0.4
1959	58.7	223	95,063	0.7
1960	65.4	235	98,008	0.9
1961	72.9	252	102,309	0.4
1962	79.4	274	104,673	0.3
1963	83.4	286	105,003	0.1
1964	86.1	296	105,059	0.1
1965	89.2	314	104,870	0.4

Sources: Civil Aeronautics Board, *Handbook of Airline Statistics, 1965 Edition* (1966); Civil Aeronautics Administration, *Statistical Handbook of Aviation* (1958); *World Airline Record, 1950–51 Edition* (Chicago: Roy R. Roadcap, 1950); U.S. Bureau of the Census, *Historical Statistics of the United States, Colonial Times to 1957* (1960).

a. Based on time from takeoff to landing, or "wheels-off wheels-on" speed.

depreciation of flight equipment. The costs of ground operations and maintenance and depreciation of ground facilities are not included.

Operating costs per seat-mile in 1954 dollars have declined from over 10 cents for the typical plane in the 1932 fleet to about 1.5 cents for the typical plane in the 1965 fleet.[6] These costs apparently move discontinuously, tracing a scalloped path through time. In the late 1920s, operating cost reductions seem to have come from the development of multiengine transports, not from the further improvement of single-engine planes. The Ford 4-AT, Fokker F-10A, Boeing 80A, Stinson SM-6000A, and Curtiss Condor had significantly lower seat-mile costs than did contemporary single-engine planes. In the early years, the costs of the first group of multiengine models fell, but later models showed little further reduction.

The next generation of planes was the all-metal, low-wing, two-engine monoplane, with retractable landing gear and other drag-reducing innovations—the Boeing 247, the Douglas DC-2 and DC-3, and the Lockheed L-10, in particular. As these were introduced, costs per seat-mile again fell fairly rapidly. But since the appearance of the DC-3, no further cost reductions have been associated with newer two-engine planes. Indeed, in many instances since World War II operating costs were lower

---

Miller and David Sawers, *The Technical Development of Modern Aviation* (London: Routledge and Kegan Paul, 1968). These were adjusted to 1954 dollars by use of the GNP implicit deflator. A combined cross-section, time-series regression was run for piston-engine planes, using as independent variables gross weight per seat, wing load, power load, model age, and dummy variables for long range or short range and for early trimotor aircraft and single-engine aircraft. With sixty-eight observations, all variables except power load are significant at the 5 percent level; $\bar{R}^2 = 0.91$. Costs for aircraft not in the sample were estimated by using in the regression equation the value for the variables that are relevant to them. An analogous estimating procedure could not be used for turboprop and jet aircraft because not enough data were available. The costs shown in Table 5-4 are, for each type, an estimate based on a trend line value, with the influence of time being the only explicit variable considered.

6. Several factors limit the precision of such estimates. Aircraft operating costs vary considerably with flight stage length and model age. Because of continued learning after planes are introduced into use and because of cost-reducing retrofits, including higher-density seating, operating costs per seat-mile tend to fall over time for nearly all aircraft studied. Thus, for example, Boeing 707 operating costs were much lower in 1965 than in 1959. For any given plane, operating costs per seat-mile tend to fall as flight stage length increases, up to the point where additional stage length requires reducing the number of seats to accommodate more fuel. The flight stage-length effect is not explicitly accounted for in the statistical estimates.

Table 5-4. Estimated Operating Costs per Seat-Mile for Aircraft Used by Scheduled Domestic Air Carriers, 1926–65

In cents per seat-mile, 1954 dollars

Manufacturer and model	Year first available for service	Operating costs, including depreciation, in first year
Ford 4-AT	1926	9.50
Fairchild 71	1928	12.82
Fokker Super Universal	1928	13.40
Fokker F-10A	1928	11.80
Hamilton Silver Streak	1928	12.13
Bellanca Pacemaker	1929	12.80
Boeing 40-B4	1929	29.14[a]
Boeing 80A	1929	10.81
Curtiss Condor Transport	1929	9.75
Fokker F-14	1929	17.24
Ford 5-AT-C	1929	7.92
Lockheed Wasp Vega	1929	10.28
Metal Aircraft Flamingo	1929	11.71
Ryan Brougham	1929	12.22
Stinson SM-6B	1929	10.89
Travel Air 6000B	1929	12.40
Boeing 221	1930	20.10[a]
Consolidated Fleetster	1930	12.10
Curtiss-Robertson Kingbird D-2	1930	9.39
Fairchild Pilgrim 100	1930	12.19
Fokker F-32	1930	6.55
Northrop Alpha	1930	10.84
Stinson SM-6000A	1930	8.31
Lockheed Orion	1931	10.90
Stinson U	1932	9.43
Boeing 247	1933	7.78
Curtiss Condor T-32	1933	7.30
Douglas DC-2	1934	6.81
Lockheed L-10	1934	4.70
Stinson A	1934	7.92
Douglas DC-3	1936	3.28
Lockheed L-12	1936	5.95
Lockheed L-14	1937	4.77
Boeing 307	1940	3.22
Lockheed L-18	1940	4.74

132

Table 5-4 *(continued)*

Manufacturer and model	Year first available for service	Operating costs, including depreciation, in first year
Douglas DC-4	1946	2.35
Lockheed L-049	1946	2.84
Douglas DC-6	1947	2.17
Lockheed L-749	1947	2.51
Martin 202	1947	2.53
Convair 240	1948	2.51
Boeing 377	1949	2.44
Douglas DC-6B	1951	1.99
Lockheed L-1049	1951	1.84
Martin 404	1951	2.38
Convair 340	1952	2.58
Douglas DC-7A	1953	1.80
Lockheed L-1049C	1953	1.86
Vickers Viscount V-745	1955	1.62
Convair 440	1956	2.22
Douglas DC-7C	1956	2.31
Lockheed L-1649	1957	2.39
Boeing 707-120	1959[b]	1.70
Douglas DC-8	1959	1.41
Lockheed L-188	1959	2.10
Boeing 720	1960	1.54
Convair 880	1960	1.73
Boeing 720B	1961	1.43
Sud Caravelle SE-210	1961	1.66
Convair 990	1962	1.55
Boeing 727	1964	1.14
British Aircraft BAC-111	1965	1.55
Douglas DC-9	1965	1.55[c]

Source: See note 5.

a. Cost estimate is high because aircraft was used primarily for mail, to which no costs are allocated.

b. The 707-120 was introduced into domestic service on December 10, 1958. In the cost estimates, 1959 is taken as its first model year.

c. Based on actual costs covering only two weeks of operation in December 1965.

for the DC-3 than for any newer aircraft with comparable performance capabilities. Cost reduction within this technology had largely ceased by World War II. Even experienced manufacturers failed in several efforts to produce improved versions. Douglas's DC-5 (1940), DC-8 (a piston-engine designation of 1946, not the DC-8 jet of the late 1950s), and Super DC-3 (1947) were designed to replace the DC-3 for short-range use, but none were commercially successful.

The four-engine, long-range transports of the late 1940s and the 1950s show the same cost pattern even more clearly. Progression from the DC-4 to the DC-6, DC-6B, and DC-7 and from the Lockheed L-049 to the L-1049 Super Constellation brought lower operating costs, but the DC-7C and the Lockheed L-1049C, L-1049G, and L-1649 Super Constellations appear to have had higher costs per seat-mile than the immediately preceding types.[7] After the piston transports, cost reductions came from a basic new technology, with first the long-range and then the short-range and intermediate-range jet aircraft. One might guess that cost reductions in jet aircraft would also tend to disappear over time.

Reductions in operating costs have generally been reflected in declining passenger fares.[8] The fare per passenger-mile (in constant dollars) on domestic trunk airlines fell from approximately 12.5 cents in 1932 to 5.5 cents in 1965, as shown in Table 5-5. One effect of the lower prices made possible by new technologies was an increase in the amount of air travel demanded.

Estimates of the magnitude of the price effect have seldom been published. Miller reports a number of such studies, but the outstanding characteristic of past studies is their lack of agreement.[9] Based on year-to-year changes in passenger-miles and in fares (unadjusted for price level changes) and with the effects of income, recent accidents, and the introduction of new types of aircraft removed, the price elasticity was estimated as −1.23 from 1932 to 1965. For the period 1947–65, a price elasticity of −1.37 was estimated.[10] Because annual observations were

7. See Aaron J. Gellman, "The Effect of Regulation on Aircraft Choice" (Ph.D. thesis, Massachusetts Institute of Technology, 1968), pp. 387–403.

8. See Joseph A. Schumpeter, *Capitalism, Socialism, and Democracy* (Harper, 1950), pp. 92–93.

9. Ronald E. Miller, *Domestic Airline Efficiency: An Application of Linear Programming* (MIT Press, 1963), pp. 18–22.

10. The regressions on which these are based are a part of the author's book mentioned in note 5. They constitute one of a set of regressions in what amounts to a block recursive system. There seem to be no identification problems arising from hidden supply effects in the demand elasticity estimation. The regression for

Table 5-5. Revenue per Passenger-Mile and Total Passenger-Miles, Scheduled Domestic Air Carriers, 1932–65

Year	Revenue per passenger-mile		Total revenue passenger-miles (thousands)
	Current dollars	1957–59 dollars[a]	
1932	0.0610	0.1282	127
1933	0.0610	0.1353	174
1934	0.0590	0.1266	189
1935	0.0570	0.1154	281
1936	0.0570	0.1145	391
1937	0.0560	0.1107	410
1938	0.0518	0.1016	480
1939	0.0510	0.1024	683
1940	0.0507	0.1024	1,052
1941	0.0504	0.0984	1,384
1942	0.0527	0.0946	1,417
1943	0.0535	0.0964	1,632
1944	0.0534	0.0962	2,177
1945	0.0495	0.0894	3,360
1946	0.0463	0.0794	5,945
1947	0.0505	0.0785	6,105
1948	0.0576	0.0804	5,997
1949	0.0578	0.0751	6,768
1950	0.0556	0.0704	8,029
1951	0.0561	0.0668	10,590
1952	0.0557	0.0622	12,559
1953	0.0546	0.0593	14,794
1954	0.0541	0.0596	16,802
1955	0.0536	0.0598	19,852
1956	0.0533	0.0584	22,399
1957	0.0531	0.0550	25,379
1958	0.0564	0.0566	25,375
1959	0.0588	0.0566	29,308
1960	0.0609	0.0587	30,557
1961	0.0628	0.0598	31,062
1962	0.0645	0.0602	33,623
1963	0.0617	0.0572	38,457
1964	0.0612	0.0560	44,141
1965	0.0606	0.0545	51,887

Source: Civil Aeronautics Board, *Handbook of Airline Statistics, 1965 Edition* (1966), pp. 47, 105.
a. Except for the years 1932–34, for which the aggregate consumer price index was used, deflation of revenue data was based on the index for all transportation.

used in the regression, these are estimates of a sort of short-term price elasticity. The range of price relatives was small, again because annual price relatives were used as an independent variable. The effects of sales promotion activities accompanying price changes are not separated, and therefore the regression coefficients do not measure pure price effects. But, to return to the impact of technological change on the market for air transportation, the estimates are good enough to make clear the relative importance of cost (price) reduction on demand. If the real, long-term price elasticity—which is presumably higher than the short-term elasticity—has been, say, -1.5, the price decreases would by themselves have accounted for only a minute part of the increase in air travel.

The changes in technology have had effects on demand other than those arising from price decreases. Improved speed, safety, and comfort have also affected it. Carriers have recognized that adopting new planes and new versions of old planes tends to stimulate demand. Indeed, in some cases new versions of aircraft have been used because of their demand-creating effect, even when the new plane had operating costs higher than the aircraft it was replacing.[11] In part, this only reflects either shifts in demand on parallel routes from one carrier to another or sales promotion activities that accompany the introduction of new aircraft. But in addition some real effect on demand does seem to occur when new types of planes are introduced. In the passenger-mile demand regression, a significantly positive regression coefficient appeared for a dummy variable, taking a unitary value for 1933, 1934, 1936, 1946, and 1959 and a zero value for the other years.[12] Such basically new aircraft as the Boeing 247, the Douglas DC-2, DC-3, and DC-4, the Lockheed L-049 Constellation, the Boeing 707, and the Douglas DC-8 were introduced in these years, and the regression shows that passenger demand tended to rise more than would have been expected from the effects of price and income alone. In the regression for the years 1947–65, the annual rate of increase in passenger-miles declines about one percentage point for each additional year in the weighted average

---

the 1932–65 period includes some arbitrarily estimated variables for exogenous events, such as the canceling of mail contracts in 1934 and the requisitioning of civilian aircraft by the government in World War II. Considerable caution is required in the use of the estimates.

11. See Gellman, "Effect of Regulation."

12. This is based on the regression covering the period 1932–65 described in note 10. The same caveats apply.

version age[13] of the aircraft added each year to the domestic trunk-line fleets. That is, when the aircraft being added were relatively new versions, demand increased more than when the planes being added, though themselves new, were older versions.

Carriers have also acknowledged—somewhat less openly—that aviation accidents influence demand. Ticket sales decrease and cancellations and "no-shows" increase immediately after an accident that receives wide news coverage. The precise effects of accidents over the years are hard to discern. A quantitatively very small, but statistically significant, negative relation appears between the annual percentage increase in passenger-miles in a given year and the sum of passenger fatalities in the previous two years.[14] The contributions of technology to increased safety and public confidence in air travel have probably had a more significant effect on demand than has price. In this, regulation played a major role, as is shown below.

Entry, Excess Capacity, Technological Change, and Regulation

Economic discussions of domestic air transport seldom emphasize the problems that arose from free entry and excess capacity in the years before the Civil Aeronautics Act of 1938. Nor do they stress the importance of technological change in altering the minimum efficient size of air carriers and in creating additional capacity.

The Industry before 1938

Summary data on the operations of scheduled domestic carriers from 1926 to 1938 are given in Table 5-6. By 1928—perhaps the first year in which passenger service became an important concern for a majority of the operators[15]—thirty-one different companies were engaged in sched-

13. To avoid the appearance of zeros in regressions which might require log transformations, version age was taken as "one" in the first year of use. Thus, the version age of the DC-6B was one in 1951, two in 1952, and so on. This variable is significant only at the 10 percent level.

14. Based again on the regression for the years 1932–65, mentioned in note 10. Interestingly, the variance in fatality rates per hundred million miles, by itself, does not help to explain the variance in passenger-mile demand.

15. Histories of the industry are alike in ascribing great importance to Lind-

Table 5-6. Summary Data on Operations of Scheduled Domestic Air Carriers, 1926–38

Year	Carriers operating during year	Carrier entries, 1926 through given year	Aircraft in use, end of year	Passenger load factor	Aircraft utilization factor[a]
1926	13	12	n.a.	12.0[b]	n.a.
1927	16	19	n.a.	26.0[b]	n.a.
1928	31	42	268	42.0[b]	5.9
1929	34	64	442	46.0[b]	7.2
1930	38	79	497	53.0[b]	7.8
1931	35	93	490	42.0[b]	9.9
1932	29	102	456	42.0	10.0
1933	24	108	418	46.7	11.0
1934	22	114	423	51.4	8.8
1935	23	118	363	48.7	11.3
1936	21	122	280	56.9	15.0
1937	17	125	291	49.1	17.4
1938	19	127	260	50.4	18.7

Sources: Table 5–9 below; Civil Aeronautics Board, *Handbook of Airline Statistics, 1965 Edition*, pp. 33, 50; *Aviation Facts and Figures* (Aircraft Industries Association of America, 1955), pp. 63, 65; *Aircraft Year Book* (Aeronautical Chamber of Commerce of America, 1929), p. 371; Civil Aeronautics Administration, *Statistical Handbook of Civil Aviation* (1958), p. 68; *World Airline Record, 1950–51 Edition*, p. 15. Unavailability of statistics is denoted by n.a.

a. Percentage of total time spent in revenue flying-hours. The decline from 1933 to 1934 is attributable to the cancellation of more contracts in that year.

b. Estimated by dividing average passengers per aircraft flight by an assumed number of seats per aircraft. Average passengers per flight is computed as revenue passenger-miles divided by revenue aircraft-hours. Numbers of seats per plane, 1926–31, were assumed to be 2, 2, 3, 4, 5, and 6, respectively.

uled domestic operations at some time during the year.[16] These operators had available for use 268 aircraft, each of which (based on 1932 data) was flown for only one or two hours a day. The passenger load factor—passenger-miles demanded as a percentage of seat-miles supplied—was about 42 percent. The total routes of these carriers measured less than 16,000 miles and daily revenue plane-miles less than 29,000.[17] On the

bergh's transatlantic flight in generating passenger demand. Total revenue and non-revenue passenger-miles rose from perhaps 3 million in 1927 to 13 million in 1928, to 41 million in 1929 (*Aviation Facts and Figures, 1945* [McGraw-Hill, 1945], p. 67). The possible effect of use by the carriers of a variety of multiengine, all-metal aircraft rather than the older, single-engine, frame-and-fabric, open-cockpit biplanes is rarely mentioned.

16. The count includes companies that operated scheduled services at any time during the year. Because of short-lived operations, this number exceeds the number of companies operating at any given time. Table 5-9 (pp. 161–65) gives detailed information on entries by year from 1926 to 1938.

17. *Aviation Facts and Figures, 1945*, pp. 66–68.

average, less than one round trip a day was flown between cities connected by scheduled carriers. By the end of 1928, at least fifteen carriers had entered and subsequently abandoned scheduled service operations. Although details are lacking, few of the carriers could have had positive returns from passenger service.

From a social point of view, the structure of the industry in 1928 was inefficient. If route expansion and higher-density service on existing routes had come from increasing the route and passenger mileage of existing operators with their existing fleets, rather than from the entry of new carriers with additional planes, the incremental costs would certainly have been lower. While more maintenance time undoubtedly was required by planes of 1928 than is needed today[18] and while overall traffic in 1928 was so thin that gains from the simultaneous and coordinated scheduling of aircraft over several routes may have been small, unquestionably routes were too short and companies were too small for the fleets to be used efficiently.[19] Yet, as is shown in Tables 5-6 and 5-9, twenty-two additional carriers entered scheduled operations during 1929. In the same year, 174 aircraft were added.

The influences of free entry and rapid technological change compounded the problem of erecting and maintaining an efficient structure. Had aviation technology remained static and had passenger demand risen as it did (though it is most improbable that the two could have occurred in conjunction), it is arguable that competitive forces would soon have effected a stable and efficient structure. But aviation technology was not static. Each year brought new carriers—despite unprofitable operations—and larger planes capable of flying faster, over longer ranges, and for more hours a year. Each new type of aircraft added to the passenger-mile capacity per plane; route extensions added to the

18. For comparison, some passenger jet aircraft are used in revenue operations for over 4,700 hours a year. See *Aviation Week and Space Technology*, Vol. 87 (Nov. 13, 1967), pp. 41 and 56, where it is indicated that Continental's fleet of twelve to thirteen Boeing 707s operated nearly 30,000 revenue hours in the first half of 1967.

19. Passenger equipment interchanges among carriers in lieu of mergers or one-carrier route extensions were used from 1940 on, with Civil Aeronautics Authority (now the Board) approval, as a means of increasing fleet efficiency. See CAB, *Handbook of Airline Statistics, 1965 Edition* (1966), p. 481. Some carriers are now trying to increase fleet efficiency by having the aircraft that are used during peak passenger periods of the day rapidly convertible to all-cargo operations in off-peak hours. See *Aviation Week and Space Technology*, Vol. 88 (Jan. 1, 1968), p. 28.

gains to be achieved from coordinate scheduling, coordinate plane assignments, and common maintenance and repair facilities.

Through 1929, the federal government accentuated, if anything, the problem of entry and excess capacity. Most of the air mail routes were awarded to small, local service carriers.[20] The government failed, at the time the Air Mail and Air Commerce Acts were being considered, to recognize the efficiency problems inherent in alternative industry structures. After 1928, certain private interests became keenly aware of these problems and acted to change public policy accordingly.

The basic structure of the domestic airline industry was virtually unaffected by regulation, having been cast in a regulatory vacuum. The Aviation Corporation, formed in 1929, quickly took over the routes of Robertson, Braniff, Continental, Northern, Central, Canadian Colonial, Colonial Western, Texas, Gulf, Embry-Riddle, and Interstate. This group became American Airways in 1930, and in 1934, American Airlines. Eastern Air Transport developed from Pitcairn Aviation and Florida Airways and was an affiliate company of North American Aviation Corporation. Its routes connected cities along the eastern seaboard. Transcontinental and Western Air (later Trans World Airlines) was a combination of Transcontinental Air Transport, Maddux Air Lines, Colorado Airways, Western Air Express (in part), and West Coast Air Transport. United's routes were composed of those of Ford, Varney, Stout, Boeing Air Transport, Pacific Air Transport, and National Air Transport.[21]

The consolidations were not in themselves adequate to control structure. Even as they were taking place, fresh entrants were appearing, many with the hope of being awarded a mail contract. By the end of 1930, such hopes were diminished. The major carriers, with the cooperation of the Post Office Department, in that year obtained passage of an amendment to the Air Mail Act, known as the McNary-Watres Act, which clearly favored larger lines in mail contract awards. This law cre-

20. R. E. G. Davies, *A History of the World's Airlines* (Oxford University Press, 1964), pp. 44–55, 123–30. Two exceptions occurred in 1927, when Boeing Air Transport was awarded the San Francisco–Chicago portion of a transcontinental mail route and National Air Transport the Chicago–New York portion. Boeing apparently recognized that lower costs might result from operating longer routes. Its bid was $2.89 a pound; the next lowest bid was $5.09. Boeing built the B-40A, a redesign of an earlier B-40, specifically for this route.

21. Ibid., pp. 44–55. Some of the lines involved in the combinations have names similar to those of subsequently organized carriers. The latter were not spin-offs from the Big Four but rather either second starts by people from companies in the original combinations (for example, Robertson, Braniff, Varney) or other lines that happened later to use the same names (such as Continental).

ated effective government control over entry and encouraged additional consolidations among the carriers then operating.[22] As Table 5-6 shows, entry declined after 1931, and because of failures and consolidations, the number of carriers discontinuing operation in succeeding years exceeded the number of new entries.

The administration of the McNary-Watres Act by Postmaster General Walter F. Brown resulted in the appointment in 1933 of a Special Committee on Investigation of the Air Mail and Ocean Mail Contracts, under the chairmanship of Senator Hugo L. Black. The favoritism shown to the major carriers was the principal reason for the brief, disastrous period in early 1934 when mail contracts were suspended and the army carried the mail. The ensuing Air Mail Act of 1934 required that carriers be separate from aircraft manufacturing companies. In addition, new methods of bidding for, and operating under, mail contracts were established. The separation of carriers and manufacturers probably caused the carriers to adopt the new aircraft types more quickly than they otherwise would have. But the act did little else either to shape the industry according to the guidelines laid down by technology or to encourage or discourage technological change in the industry. In accordance with the provisions of the act, a federal aviation commission was established by President Roosevelt to study and recommend federal regulatory policy. De facto private regulation of passenger service continued largely unabated.

Whatever the shortcomings of policy during this period, the structure of the industry did change in a direction consonant with increasing efficiency. As lines were consolidated, routes were extended, and new aircraft were adopted, passenger traffic rose rapidly. Scheduled domestic operations amounted to 127 million passenger-miles in 1932 and to 480 million in 1938. In 1932, 304 million seat-miles of scheduled service were supplied, with 456 aircraft in operation at the end of the year. In 1938, more than three times that many seat-miles were supplied, with only slightly over half as many aircraft.

It is easy to misjudge even this response to the demands of structural efficiency. Table 5-7 shows that of the fifteen domestic trunk air carriers operating at the end of 1938 eight had fewer than eight aircraft in their fleets.[23] Only seven DC-2s and three DC-3s were in the fleets of lines

22. Ibid., pp. 123–24.
23. Sixteen domestic trunk carriers were certified in 1938. Marquette Airlines, however, apparently was not in passenger operations at the end of the year and is not included here. It owned three Stinson trimotors. Its routes may have been operated by Transcontinental and Western Air, with which it merged in 1941.

Table 5-7. Aircraft Operated by Domestic Trunk Air Carriers, December 31, 1938

	Aircraft manufacturer and model							
	Douglas		*Boeing*	*Lockheed*			*Stinson tri-motor*	*Total in fleet*
Carrier	*DC-2*	*DC-3*	*247*	*L-10*	*L-12*	*L-14*		
Big Four								
American	15	30	—	—	—	—	—	45
Eastern	10	10	—	—	—	—	—	20
Transcontinental and Western Air	14	19	—	—	—	—	—	33
United	—	35	17	—	—	—	—	52
Other trunk								
Braniff	7	—	—	6	—	—	—	13
Chicago and Southern	—	—	—	5	—	—	—	5
Continental	—	—	—	—	3	—	—	3
Delta	—	—	—	5	—	—	—	5
Inland	—	—	6	—	—	—	—	6
Mid-Continent	—	—	—	4	—	—	—	4
National	—	—	—	3	—	—	2	5
Northeast	—	—	—	3	—	—	2	5
Northwest	—	—	—	7	—	9	—	16
Pennsylvania-Central	—	—	12	—	—	—	—	12
Western	—	3	4	—	—	—	—	7
Total, all carriers	46	97	39	33	3	9	4	231

Source: Civil Aeronautics Board, *Annual Airline Statistics: Domestic Carriers, Calendar Years 1938–42* (1943), pp. 133–34.

other than the Big Four trunk operators. The Lockheed L-10 Electra, a ten-passenger aircraft with relatively low seat-mile costs and, for loads of ten or fewer passengers, much lower total passenger-mile costs than the DC-2 or DC-3, was the most popular plane among the smaller carriers.

It is likely that substantial excess capacity remained. The 97 DC-3s and 33 Lockheed L-10s could have been used for at least 2,400 hours a year without increasing maintenance costs or threatening safe operations. In the acute shortage years of 1943 and 1944, when the domestic trunk fleet consisted primarily of DC-3s, hours of use averaged nearly 3,700. Assuming a block speed of 150 miles an hour, with twenty-one seats for the DC-3s and ten seats for the L-10s, these 130 planes alone could have provided a technical seat-mile capacity nearly as great as the number of seat-miles actually obtained from the 231 planes in the entire trunk carrier fleet. These data suggest that further consolidations of carriers and extensions of routes by existing carriers would have been in order. The relevant costs of operating in the territories of smaller carriers and in new areas would, to the extent that excess aircraft capacity was used, have been the out-of-pocket costs of more intensive overall fleet operation rather than the full costs of expanding the underutilized fleets of the individual lines.

The Industry after 1938

The Civil Aeronautics Act of 1938 was not the result of public dissatisfaction with the rates or services of private carriers. Rates had generally been falling, service was being extended into new areas, frequency of service between major cities was increasing, and new types of aircraft were being used. Virtually all planes in use by the trunk carriers were all-metal, low-wing, two-engine monoplanes—modern and sleek by the standards of the day, and indeed for years thereafter.

The 1938 act was partly the result of the unfavorable publicity given to the administration of Postmaster General Brown and his so-called spoils conference method of dividing markets among the Big Four. But the effects of this would probably have subsided without further legislation had there not been a rash of highly publicized fatal accidents. While eight passengers died in scheduled domestic air operations in 1933, deaths were seventeen, fifteen, forty-four, and forty, respectively, in the years from 1934 through 1937.

The death of Senator Bronson Cutting in the crash of a Transcontinental and Western Air DC-2 in 1935 precipitated a bitter controversy about safety, the behavior of the carriers, and the responsibility of government. The investigation of this accident revealed that:

1. The pilot had not taken the required quarterly medical examination.

2. The pilot, after an absence of over six months, had not obtained renewed approval to fly in the division of the line in which the crash occurred.

3. The copilot did not hold a scheduled air transport pilot rating.

4. The TWA dispatcher at Kansas City should have grounded the flight at Wichita because of Kansas City weather and the limited fuel reserve of the aircraft.

5. The aircraft was cleared for instrument flying although its two-way

radio was not operating properly for transmitting at the night frequency, a condition the pilot was aware of.

6. The Weather Bureau had failed to forecast the hazardous weather conditions.

7. The radio range beacons of the Bureau of Air Commerce, while no worse at the stations related to the crash than elsewhere, were inadequate, especially in poor weather.[24]

Other aspects of industry performance besides safety quite obviously lay behind the 1938 act. In addition, the act seems to have stemmed from industry reactions to the tendency for free entry to generate chronic excess capacity.[25] While entry consistent with the act's purposes was provided for in the legislation, no new trunk carriers have been certified since the period immediately following its passage.[26] Mergers and suspensions, on the other hand, have caused the disappearance of some of the sixteen trunk carriers operating prior to passage of the act (protected by a grandfather provision) and of the three other carriers originally certificated.

For the most part, regulation has prevented mergers among larger carriers and among carriers serving the same areas. Because the smaller carriers operate largely unrelated lines and hence have not selected one another as merger partners, the attention given by the Civil Aeronautics Board (CAB) to the size of the Big Four and to parallel areas of service has probably prevented some mergers that, when they were proposed, would have brought about route extensions and conceivably some operating economies. The first decision under the act involved a proposed consolidation of United Air Lines and Western Air Express, in which such conditions prevailed. Except where some sort of "failing firm" doctrine could be applied, permission for mergers among carriers serving the same city-pairs has been denied.[27]

24. The political controversies of the period seem to have obscured the underlying importance of safety in the 1934–38 discussions. See, however, D. R. Whitnah, *Safer Skyways: Federal Control of Aviation, 1926–1966* (Iowa State University Press, 1967), who discusses the Cutting crash on pp. 116–18.

25. For a near contemporary explanation of efforts by the industry to obtain passage of the act, see Henry L. Smith, *Airways: The History of Commercial Aviation in the United States* (Knopf, 1942; Russell and Russell, 1965), pp. 94–102.

26. See Lucile Keyes, *Federal Control of Entry into Air Transportation* (Harvard University Press, 1951), and "A Reconsideration of Federal Control of Entry into Air Transportation," *Journal of Air Law and Commerce,* Vol. 22 (Spring 1955).

27. Something akin to failing firm arguments, together with "service improve-

The extension of scheduled air transportation into new areas—regarded here as being a qualitative as well as a quantitative dimension of the industry's product—has come in spurts. One was from 1936 through 1941, the years immediately before and after the 1938 act. These extensions coincide with the higher and more stable earnings in the industry that came with the introduction of the DC-3 and the settling of the structural characteristics of markets in this period. Note is made below of the extent to which carriers act as sales-maximizing oligopolists with earnings constraints. The second period of rapid route expansion was from 1944 through 1950. This was the result initially of the relaxing of World War II restraints on operations and subsequently of the original awarding of routes to local service, or feeder, lines. The third period was from 1955 to 1961, when the trunks began service to several new cities that might otherwise have been served only by feeder lines, and when increased paralleling of routes was allowed to develop.[28]

In addition to the effect on industry structure of the CAB's encouragement of feeder lines, the structure in the period immediately after World War II was affected by the entrance—legal and extralegal—of large-scale nonscheduled, or "irregular," supplementary carriers.[29] This happened at a time when the principal carriers were experiencing drastic declines in load factors, adding new planes as replacements for DC-3s and DC-4s, reporting negative current rates of return, and seeking increased passenger fares. The appearance of the "non-skeds," rather than direct action by the CAB or competition among the regular carriers, is the most important reason for the introduction by the trunk carriers of the reduced "all coach" and economy fare and economy service operations in the new aircraft used over their main route segments. Although special family and quantity-type discounts had been made before World War II, efforts to differentiate service emphasized luxury rather than economy during the early period; for example, the DC-3 was originally ordered in a sleeper configuration. By the 1950s, a series of CAB actions had removed the threat of price competition from non-skeds, leaving only nonprice competition from a small number of scheduled carriers on the longer, higher-density, and potentially most profitable routes.[30]

---

ment-cost reduction" arguments, was present in the Mid-Continent–Braniff merger (1952) and the United-Capital merger (1961). See Richard E. Caves, *Air Transport and Its Regulators* (Harvard University Press, 1962), Chap. 8.

28. Ibid., pp. 204–31, for additional detail.

29. Ibid., pp. 12–14, 91–93, 277–85, 370–71.

30. Ibid., pp. 171–76.

Table 5-8. Aircraft Operated by Domestic Trunk Air Carriers, December 31, 1958[a]

Aircraft manufacturer and model

Carrier	Douglas				Boeing	Lockheed				Convair		Martin		Vickers Viscount	Total
	DC-3	DC-4	DC-6 and DC-6B	DC-7, DC-7B, and DC-7C	377	L-18	L-649 and L-749	L-1049	L-1649	240	340/440	202	404	745 and 812	in fleet
Big Four															
American	—	—	83	58	—	—	—	—	—	54	—	—	—	—	195
Eastern	—	1	9	48	—	—	23	38	—	—	20	—	58	—	197
Trans World	—	6	—	—	—	—	71	44	29	—	—	10	37	—	197
United	1	—	91	55	—	—	—	—	—	—	54	—	—	—	201
Other trunk															
Braniff	20	—	10	6	—	—	—	—	—	—	30	—	—	—	66
Capital	19	12	—	—	—	—	11	—	—	—	—	—	—	59	101
Continental	13	—	3	5	—	—	—	—	—	—	9	—	—	14	44
Delta	12	—	7	21	—	—	4	—	—	—	28	—	—	—	72
National	—	—	12	4	—	7	—	7	—	—	18	—	—	—	48
Northeast	11	—	10	—	—	—	—	—	—	5	—	—	—	7	33
Northwest	4	12	21	17	9	—	—	—	—	—	—	—	—	—	63
Western	2	—	24	—	—	—	—	—	—	6	—	—	—	—	32
Total, all carriers	82	31	270	214	9	7	109	89	29	65	159	10	95	80	1,249

Source: Federal Aviation Agency, *Statistical Study of U.S. Civil Aircraft as of January 1, 1959* (1959), pp. 4–5.

a. Includes cargo versions of passenger types of aircraft (for example, DC-6A).

Rivalry in equipment and in the amenities of first-class air travel had existed before the 1950s, but it intensified at about this time. Aspects of the pre-World War II equipment rivalry are discussed below. Of the transcontinental carriers, American and United added, in order, Douglas DC-6, DC-6B, DC-7, and DC-7B aircraft, while Transcontinental and Western Air purchased Lockheed 649, 749, and 1049 Constellations and 1049C, 1049G, and 1649 Super Constellations. Eastern, with routes paralleling Delta, National, Northeast, and Capital in different segments of its domestic operations and Pan American in the Caribbean, had mixed fleets. Competitive promotional policies were based on differentiated equipment and service,[31] causing new transcontinental "stretched" versions to be ordered, even though higher costs were both anticipated and realized.

Despite rivalry in aircraft additions, the composition of the 1958 fleets of the domestic trunk carriers gives little impression that excess capacity comparable to that of the 1930s remained through the 1950s. Western Air Lines' fleet of thirty-two planes was the smallest (Table 5-8), yet—indicative of technical progress as well as the growth of the industry—its twenty-four DC-6 and -6B aircraft alone, with reasonable assumptions concerning seat density, block speeds, and available annual hours of service, had a seat-mile capacity more than 50 percent greater than that of the entire scheduled domestic fleet of 1938. No major line was so small that its fleet was forced to low hours of utilization through "indivisibilities" resulting from limitations on schedules and route configurations. The overall passenger load factor for 1958 was 60.0 for the trunks and 59.4 for total domestic operations, compared with 50.3 and 50.4 in 1938. Indeed, this improvement was so obvious that the CAB inaugurated a policy of increasing route overlapping—in the name of competition—beginning in 1955.[32]

Permitting and encouraging competition may not be as efficiency-inducing as at first appears. Where the product can be fairly easily differentiated, and where sellers are few and entry is restricted, nonprice competition among sellers who measure short-term success in terms of market shares may lead to an industry output mix that is not ideal from the consumer's view and to excess capacity. The small number of airlines serv-

31. See Gellman, "Effect of Regulation," especially pp. 406–502.
32. The Southwest-Northeast Service Case and the Denver Service Case are taken as turning points in the use of a competitive criterion. See Caves, *Air Transport,* pp. 204–31.

ing parallel routes, with the price-competitive non-skeds blocked, may have given rise to too much expensive first-class service, the introduction of uneconomical new kinds of aircraft, and the peaking of service by all carriers in the popular, high-traffic hours of the day.[33]

The introduction of jets has perhaps accelerated the trends noted for the mid-1950s; at least there is no evidence of any reversal. The best summary of the extent of technological change that came with jets is that one Super DC-8 of 1968, in a 250-passenger configuration, operating for 3,500 hours a year at an average block speed of 500 miles an hour, could, if full, produce almost as many passenger-miles in a year as the entire domestic fleet of 1938.

The Influence of Regulation on Air Carrier Research, Development, and Innovation

The preceding sections outline the nature of technological progress and its influences on the development of regulation of air commerce. This section is concerned with the sources of the new technologies and the effects of regulation on technological progress. In particular, the effects of regulation on research, development, and innovative activities of the carriers will be considered.

Despite the volume of recent economic research on the interrelations of technological change, market conduct, and market structure, the fact remains that little is known about them. Because of this, certain aspects of technological change which were relatively unaffected by regulation are suggested first. Next, a few aspects of change where regulation quite obviously did have an effect are noted. Finally, some aspects on which regulation may or may not have had significant effects are discussed.

The Basic Technologies for Aircraft Engines and Frames

Up to the late 1920s, the potential for mail and passenger carriage by air significantly influenced the development of new airframes. Aircraft such as Boeing's 40, 80, 220, and 221 and the Ford Tri-Motor were developed for such purposes following the 1925 legislation, and the later of these aircraft were the first in the United States to make use of a number

33. This argument is essentially that of Peter O. Steiner in "Program Patterns and Preferences, and the Workability of Competition in Radio Broadcasting," *Quarterly Journal of Economics*, Vol. 66 (May 1952), pp. 194–223.

of new developments in technology. Among the latter were thick-root, tapered-planform, cantilever wings; all-metal, stressed-skin construction; and multiple engines. While American firms such as Stout, Boeing, and Douglas worked with these technologies in the 1920s, primary credit for their basic development rests with the European companies, especially Junkers and Fokker.[34]

After the very early 1930s, most improvements resulted from efforts to get better planes for military use. This motivated the development in the United States of better radial engines, on which the improved aircraft depended, and of airfoils, airframes, and aircraft engines used in commercial planes since the early 1930s. Improvements in radial engines, Schlaifer has shown, came exclusively from funds provided by the government in the early 1920s for predominantly military objectives. The technology of turbojet and turbofan engines was also developed for reasons largely unrelated to commercial aviation.[35] Commercial aircraft after 1930 were based almost entirely on underlying, fundamental technologies developed at government expense by private industry, by government research organizations, and especially by the highly influential National Advisory Committee for Aeronautics (NACA)—the predecessor of the National Aeronautics and Space Administration (NASA). The results of this research were used to develop both transports and bombers in the early 1930s. After 1935, bombers tended to take precedence over transport planes. The few attempts at basic advances in commercial aircraft without antecedent military planes were commercial failures.

New commercial aircraft have required substantial development work specific to the commercial purpose. Quite obviously, vast resources have been spent to that end,[36] and regulation has affected the amounts thus spent and the results. The separation of aircraft manufacturing from air transportation in 1934 seems to have stimulated the development and acquisition of aircraft with lower operating costs and greater traffic-generating characteristics. Nevertheless, since 1930 neither regulation nor the commercial interests of the air carriers have been of appreciable im-

34. J. L. Nayler and E. Ower, *Aviation: Its Technical Development* (Dufour, 1965), pp. 101–06, 271–75. That the carriage of mail was considered a dominant feature of U.S. air traffic in the late 1920s is shown by the fact that the aircraft were called "mail planes," not "transports."

35. Robert Schlaifer, *Development of Aircraft Engines,* and S. D. Heron, *Development of Aviation Fuels* (in one volume; Harvard University, Graduate School of Business Administration, 1950), pp. 156–98, 332–508.

36. Douglas, for example, is reported to have spent $215 million to develop the DC-8 alone. See *Wall Street Journal* (Sept. 11, 1959), p. 23.

portance in expanding the basic technological frame within which the most successful of the commercial aircraft have been developed. Commercial aircraft manufacturers have had a rapidly advancing body of technical knowledge available to them at rather low cost because of government-sponsored research.

The Boeing 247—the first modern airliner—was introduced into service in 1933. An all-metal, low-wing, stressed-skin, two-engine monoplane with retractable landing gear, in its initial configuration it had a gross weight of 12,650 pounds and could carry ten passengers at a cruise speed of about 155 miles an hour over an effective full-load range of nearly 500 miles. The 247D cruised at over 180 miles an hour and had a somewhat longer full-load range. The DC-2, of 1934, weighed 18,200 pounds and carried fourteen passengers at roughly the same speed over the same range. These were the advanced commercial aircraft of the day.

The all-metal, two-engine monoplane bomber was developed a short time before the 247 or the DC-2. The Martin 123—later to become the B-10 bomber—first flew in 1931. As with the later DC-2 and 247, this plane had retractable gear,[37] was all metal except for fabric-covered control surfaces, had the new drag-reducing NACA cowlings over the engines, and had a top speed of about 200 miles an hour. Its gross weight in the experimental 123 version was 12,560 pounds; in the B-10 version, 16,400 pounds. Its bomb load of 2,000 pounds was at least equivalent to the ten-passenger load of the 247.[38] The 247, the DC-2, and contemporary military planes, revolutionary compared to the aircraft they replaced, were well within the bounds of basic technical developments achieved by NACA and other research agencies. The DC-1, a true prototype of the DC-2, is reported to have been designed in five days.[39]

The conjunction of military and commercial developments in the late 1920s and early 1930s is illustrated by Boeing projects. The Model 200 Monomail provided aerodynamic and structural advances that helped in

37. The gear of the DC-2 did not retract fully. This fact made failure of the mechanism less hazardous, since in belly landings the partially protruding wheels rather than the fuselage or engine pods made contact with the ground. Retractable landing gears had appeared on racing aircraft as early as 1920. See Nayler and Ower, *Aviation*, pp. 72, 255.

38. Ibid., pp. 88–89.

39. Smith, *Airways*, p. 340. The five days covered what would now be called "program definition." The design was specified in the contract with Transcontinental and Western Air, but a great deal of engineering development remained to be done.

the development of Models 214 and 215, which became the B-9 bomber. Models 220 and 221, variants of the 200, led to Model 246, the improved B-9A bomber. From the latter, the Boeing 247 emerged.

The reliance of manufacturers on research done outside the industry is indicated by the following brief histories of the technological features of the aircraft. Improved radial engines came from government funds and indeed from government research, especially at McCook Field. The fairing of engines into leading edges of wings, which had significant drag-reducing effects, was the result of NACA research. The wing sections of the DC-1, -2, and -3, while used previously on the Northrop Alpha and Gamma, were the NACA 2215 section at the root and the NACA 2209 section at the tip. The development of trailing-edge flaps, used on some foreign aircraft since 1914, benefited materially from NACA wind tunnel research. Most of the aircraft of the early 1930s used the famous NACA long-chord cowling. Monocoque fuselage and stressed-skin cantilever wing construction had been developed in Europe some years before and had been used previously in the United States in such planes as the Lockheed Vega, the Boeing XP-9 fighter, the Boeing Models 200 and 221, and a variety of racing aircraft. Thus the new planes resulted, in a sense, from a happy conglomeration of a large number of fairly well-known aerodynamic principles and construction practices.

The DC-3 was a serendipitous "stretch" of the DC-2. Yet even as it was being introduced in 1935, the carriers provided joint financial support to Douglas for a four-engine transport capable of carrying at least forty passengers over a longer range than that of the DC-3.[40] This was pressing on the limits of the then-existing frame of technology. So too was the specification issued in the same year by the Army Air Corps for an experimental bomber with approximately the same flying characteristics.

The DC-4E, which resulted from Douglas's first years of research, had a gross weight of 65,000 pounds. Only one was built. It had three vertical tail fins, much like those of the later Constellations. The DC-4E was designed to carry forty-two passengers. Its heavy weight relative to its passenger-carrying ability and its wing loading of only 30 pounds a square foot indicate potentially high operating costs. The DC-4, built af-

40. *Aviation*, Vol. 37 (July 1938), pp. 20, 30–31. The carriers that wanted to avoid "costly unnecessary competition" were American, Eastern, Transcontinental and Western Air, United, and Pan American.

ter the DC-4E, first flew in 1942, had a gross weight of 73,000 pounds, ordinarily carried about fifty-four passengers, and had a wing loading of 50.1 pounds a square foot.

One military development roughly contemporary with that of the DC-4 was the Boeing Model 294 (the B-15 bomber). It too was generally unsuccessful, with low wing loads and poor performance. The first really successful four-engine plane was Boeing's Model 299, which in later versions was the B-17 bomber. The early B-17s had a gross weight of about 60,000 pounds, a range of 1,100 miles, a cruise speed of about 220 miles an hour, and a normal operating altitude of 25,000 feet. The Boeing 307 Stratoliner, introduced in 1940 and of about the same design as the B-17 bomber, was the first pressurized-cabin, four-engine transport in the U.S. fleet. The 307, however, weighed only 42,000 pounds and had a passenger capacity of thirty-three persons and a wing loading of 33.5 pounds.

Lockheed began development of the L-049 Constellation in about 1939, with support from Transcontinental and Western Air, which had withdrawn from the DC-4 project. The Constellation, though developed after the DC-4E, also strained existing technology. It was to weigh 98,000 pounds gross and carry fifty-one passengers for a range of over 2,000 miles in a pressurized cabin. Its wing loading was 59.3 pounds. The plane was in production for delivery to TWA at the time of U.S. entry into World War II and was delivered instead to the government for military transport use.

Research by NACA became increasingly oriented to military requirements after 1935. The DC-4 and L-049 were the last successful commercial developments that did not have military antecedents. The military and commercial planes were not always produced by the same companies; however, engineering personnel are quite mobile among companies, thus making easier the interfirm transfer of technology. The DC-6 and DC-7 series were stretches of the DC-4, although substantial stretches. Similarly, the L-649, L-749, L-1049, and L-1649 Constellations were stretches of the L-049. But even prior to the stretchings, military aircraft had pushed into and beyond the technical and performance characteristics involved. The B-29 bomber, a prototype of which flew in 1942, had a gross weight of 135,000 pounds, pressurized crew compartments, a total of 8,800 horsepower, a maximum speed in excess of 350 miles an hour, and a range of 4,100 miles. The B-50 version, deliveries of which began in 1947, had 14,000 horsepower, a top speed of more than 400 miles an hour, and a range of over 6,000 miles with 10,000 pounds of bombs. The DC-7 of 1953 weighed about 122,000 pounds, had 13,000 horsepower, a range of 3,600 miles, and a cruise speed of 350 to 370 miles an hour, and the L-1049C of the same year had characteristics very similar to those of the DC-7.

Much the same pattern emerged in the development of the first U.S. commercial jet transports. The principal characteristics of the Boeing 707 were defined fairly specifically as early as 1948, but the aircraft was not put into commercial use until 1958. The de Havilland Comet I flew in prototype in 1949; it entered commercial service with British Overseas Airways Corporation in 1952. It was a small aircraft, seating thirty-six to forty-four passengers. Its four engines had only 5,000 pounds of thrust each, and it cruised at only 490 miles an hour. The 200,000-pound, six-jet, 600-mile-an-hour, swept-wing B-47 entered service in 1950. The XB-52 and YB-52 first flew in 1952; the very similar B-52A entered service in 1954. The initial configuration had a gross weight of 350,000 pounds; subsequent versions reached 488,000 pounds. The XB-52 had eight jet engines, each of which had 8,700 pounds of thrust. Thrust increased in production versions, with the B-52F having 13,750 pounds an engine and the B-52H, 17,000 pounds an engine. The loading on the 35-degree back-swept wings of the B-52 reached over 122 pounds a square foot, and it had a top speed of 650 miles an hour and a service ceiling of over 50,000 feet. The initial versions of the Boeing 707 and the DC-8, put into use some four or five years later, were much smaller aircraft with lower wing and engine loadings. Indeed, the B-52, in terms of weight and performance, is more in the class of the C-5 military transport and the "jumbo jet" subsonic commercial transports.

There were other attempts to press on technological frontiers with commercial aircraft, which, unlike those made for the DC-4 and L-049, were never successful. Notable among these are the Consolidated CV-37 in 1946, the Republic Rainbow in the same year, and the Hughes H-4 and the Lockheed Constitution, both in 1947. Their proposed gross weights were, respectively, 320,000, 114,200, 400,000, and 184,000 pounds. The CV-37 was to carry 216 passengers; the Constitution, 180. None were ever put into service.

In sum, it seems reasonably clear that for a time before 1938, and in the years since, the basic technology for aircraft engines and frames developed largely outside commercial aviation and its regulation. The possible influences of regulation on specific uses and developments within the basic technologies are considered below.

Safety

Regulation has almost certainly had an effect on safety. Here, as in the case of engines and airframes, much of the basic research was done for military aviation and for other purposes not directly related to aviation.[41] But regulation clearly affected the use of these technologies and their specific adaptation to commercial aviation.

Research had been conducted by the carriers on questions affecting safety well before 1938. United Air Lines offers probably the best example, having established a research organization as early as 1929.[42] Communications research—particularly that relating to the use of two-way radio for ground flight control—was an important part of the work. United spent $400,000 between 1930 and 1940 on its research program.

Nonetheless, for much of the important safety apparatus, private costs and benefits to an individual carrier diverge from social costs and benefits. In particular, ground radio installations for navigation and ground-controlled approach systems, which by their nature can be used by all aircraft with appropriate on-board equipment, provide external benefits when undertaken by a single carrier. The social valuation of the cost of crashes also probably exceeds the cost as viewed by the carriers.[43] At least their behavior before 1938 suggests that, without regulation, they were unwilling to incur the costs incident to a reduction in fatality rates.

An early step taken after passage of the Civil Aeronautics Act in 1938 was to require the use of approved radio and flight equipment for safety purposes. From late 1937 through 1939, considerable attention was given by the carriers to the effective use of this equipment,[44] and the accident and fatality rates fell dramatically. From March 26, 1939, to Au-

41. For a brief summary of the historical development of navigational aids, see Nayler and Ower, *Aviation,* pp. 243–51.
42. Jay P. AuWerter, "Airline Research," *Aviation,* Vol. 39 (September 1940), p. 30.
43. Gellman ("Effect of Regulation," pp. 183–84) notes that some carriers may actually report an accounting profit from crashes.
44. For the list of approved radio equipment, see Civil Aeronautics Authority, *Air Commerce Bulletin,* Vol. 10 (Feb. 15, 1939), pp. 217–19. For articles on its use, see, for example, Henry W. Roberts, "Radio System of American Airlines," *Aero Digest,* Vol. 31 (October 1937), p. 36; T. F. Collison, "Strengthening Aviation's Wings: Aids to Safety in Operations of Eastern Air Lines," *National Safety* (June 1938), p. 21; and Charles I. Stanton, "The CAA and Aircraft Radio," *Aero Digest,* Vol. 35 (September 1939), p. 59. Many similar pieces appeared at about this time.

gust 31, 1940, no fatal accidents occurred in U.S. domestic scheduled air service. As Table 5-3 shows, the fatality rate was uniformly lower after 1938 than in previous years.

Little direct research by the carriers was related to safety. Most related instead to cost-reducing or revenue-generating activities, such as aircraft maintenance, equipment scheduling, traffic flows, passenger handling, freight and baggage handling, ticket reservation systems, food and beverage service, and in-flight passenger entertainment.[45] The exceptions were Transcontinental and Western Air, which, in association with Hughes Aircraft, did some research on radar safety devices,[46] and United, which for some time has been engaged in developing detection devices for clear air turbulence. The effects of regulation on these are uncertain.

Aircraft Selection and Specific Performance Attributes

The 1934 legislation separating aircraft manufacture from commercial air carriage affected the carriers' demand for planes.[47] Prior to this requirement, none of the larger carriers had been fully free to select what, from their point of view as operators, would have been the best available aircraft. The principal purchaser of Fokker's Super Universal F-10 and F-14 aircraft, for example, was Western Air Express, a company controlled in common with Fokker.[48] The operating costs of the F-10 and F-14 were high compared with those of other available aircraft, and they had no obvious offsetting technical or performance advantages.

The primary demand for the Boeing 40s, 80s, 221s, and 247s came from Boeing Air Transport and its successor, United. Indeed, none of these aircraft were delivered new to any carrier other than Boeing or United. Captive demand deriving from common ownership seems to explain the original use of the aircraft.

The first-hand demand for the Curtiss Condor, the Curtiss-Robertson Kingbird D-3, and the Curtiss Condor T-32 arose from complex control interlocks among the Curtiss companies, General Motors, North Ameri-

45. For an indication of the scope of activities, see G. L. Christian III, "Airlines Study New Systems," *Aviation Week,* Vol. 52 (May 29, 1950), p. 15.
46. Robert Hotz, "TWA, Hughes Aircraft Developing New Airline Radar Safety Devices," *Aviation News,* Vol. 7 (May 12, 1947), p. 28.
47. Smith, *Airways,* pp. 339–41; Caves, *Air Transport,* p. 100.
48. P. Bowers, "The American Fokkers," Pt. 3, *Journal of the American Aviation Historical Society,* Vol. 12 (Fall 1967).

can Aviation, and Eastern and American Airlines.[49] American Airlines was also the major purchaser of Stinson trimotors, ostensibly because of further links with the Aviation Corporation and other E. L. Cord interests. That Transcontinental and Western Air selected Douglas to develop the DC-1 and DC-2 may be related to the fact that North American Aviation, which controlled TWA at the time, owned 89,000 shares of Douglas.[50]

Development of the DC-1 and DC-2 marks the beginning of overt rivalry among the trunk-line carriers for aircraft that are demand-generating. In 1932 and 1933, Transcontinental and Western Air had a fleet of Fords, Fokkers, Lockheed Vegas and Orions, and Northrop Alphas. None was capable of holding passenger demand on transcontinental routes where the Boeing 247s of United could be used. The highly publicized, record-breaking flight of Jack Frye of TWA and Eddie Rickenbacker of Eastern from Los Angeles to Newark in the first DC-2 indicates the very personal nature of this rivalry.[51]

Equipment rivalry, clearly abetted by the separation of manufacturing from air transport, continued. The DC-3, which dominated the acquisitions of the trunk carriers from 1935 until World War II, was the only aircraft universally selected by all the carriers. Shortly after passage of the Civil Aeronautics Act of 1938, Transcontinental and Western Air broke away from the joint efforts to develop the DC-4. Since that time, the principal carriers have engaged in a form of market-share competition, one element of which has been differentiation in aircraft selection and, with it, differentiation in services offered to passengers.

Gellman argues that this form of competition, and hence the configurations and time-patterning of specific aircraft added to the fleets, are the result of regulation. Because regulation mitigated, if it did not eliminate, price competition and did little toward controlling nonprice competition, Gellman concludes, "There is little doubt that, in the United States, the Civil Aeronautics Board has been the single most influential force determining equipment investment decisions for the carriers under its jurisdiction."[52] Of course, the CAB has controlled forms of nonprice competi-

49. Smith, *Airways,* pp. 337–38. The ownership and control interlocks are unsympathetically explained in Elsbeth E. Freudenthal, *The Aviation Business, From Kitty Hawk to Wall Street* (Vanguard, 1940), pp. 98–133.

50. Smith, *Airways,* p. 340.

51. Ibid., pp. 251–53. This flight was also motivated by the fact that nine days earlier President Roosevelt had ordered the cancellation of mail contracts, effective the day following the flight.

52. Gellman, "Effect of Regulation," especially pp. 198–263.

tion which have the near-equivalent of a price component. For example, the board disapproved United's attempt to introduce coach service on flights using aircraft that had seating densities and other services much like those of first class.

This argument calls for very careful interpretation. It seems undeniable if it merely hypothesizes that the forms of competition countenanced by the CAB (for an industry with the structural characteristics of commercial aviation) tend to produce market-share rivalry based on product differentiation. But the extent to which the CAB shaped the structure and behavior of the carriers is a major question. The view taken here is that the CAB had very little impact on either structure or behavior. In a sense —and admittedly only when viewed in terms of long-run policy effects rather than particular decisions—the CAB has behaved much as would a reasonably far-sighted trade association operated by a group of oligopolistic carriers that had partially overlapping, but far from coincident, market areas.

Competition through equipment investment decisions began before 1938. It began as soon as a substantial volume of passenger traffic developed on a small number of carriers serving city-pairs that could effectively use aircraft embracing newer technologies. This was the period of the Boeing 247, the DC-2, the Stinson A, and the Condor T-32. Interestingly, the code worked out by these carriers under the National Industrial Recovery Act contained provisions concerning unfair and cutthroat competition that, while stated in the most general terms, would in practice have condemned price competition and permitted nonprice competition.[53] Thus self-regulation would have incorporated the features of the 1938 legislation on this point.

With the concentrated structure of the industry in the 1930s, the separation of manufacturing from airline operation in 1934 increased the extent to which the carriers could satisfactorily resolve problems arising from their oligopolistic interdependencies. With separation, the carriers became solely oriented toward air carriage, which did not necessarily coincide with the interests of the aircraft manufacturers. Concomitant with this development was the virtual termination of significant new entry after 1934.

The protection from non-skeds that regulation afforded in the 1947–50 period undoubtedly limited price competition and accordingly encouraged equipment competition for a number of years. This is quite in

53. For industry's view of the need for this and entry barriers, see *Aviation*, Vol. 33 (October 1934), p. 324.

accord with the Gellman hypothesis.[54] The longer-term effects would probably not have been very different even if the non-skeds had been able to enter more freely. The non-skeds started with surplus World War II aircraft, the supply of which was limited. As these aged, passenger demand might have slackened if newer craft had not been used as replacements. Price competition, if it had emerged in force, would soon have eliminated most of the non-skeds. While passenger-mile costs would have been lower for a carrier that operated on a truly nonscheduled basis— by waiting for a nearly full load—passenger demand would have switched to the carriers that operated more regularly and more nearly on a schedule. In this market context, the remaining irregular carriers would have developed operations closely resembling those of the regulars. In time, the irregular carriers would simply have joined the ranks of the regulars, and nonprice competition would have prevailed again. In the process of merging and behavioral coalescing, a service such as the present-day shuttle might well have emerged.

Regulation may also have forestalled the opening of the low-return routes eventually covered by local-service operators. Without governmental regulation, market-share competition among the major operators might have encouraged one or another of the lines to open feeder operations as a means of generating traffic on main routes. If one had done so, the others would have felt compelled to follow, though not necessarily in the same feeder areas. If regulation had this restricting effect, it is because the explicit profit constraint inherent in CAB regulation was more effective than the implicit profit constraint characteristic of oligopolistic market-share rivalry. But the behavior of the carriers, with regulation, was similar to that Baumol ascribes to sales-maximizing oligopolists.[55] The carriers pressed for additional routes and more passengers, although not during periods when profits were very low or into areas in which returns would be low.

The increase since 1955 in route overlapping among the trunk carriers, and between trunks and feeder operators, may again indicate that the regulator has adapted to the market conditions and values of those being regulated. In this period, for the first time, the minimum efficient size of operating units was small enough relative to the size of markets that multiple-market occupancy in other than trunk-line regions would

54. In "Effect of Regulation," Gellman covers in detail only the period 1947–58.

55. William J. Baumol, *Business Behavior, Value and Growth* (rev. ed., Harcourt, Brace and World, 1967), Chaps. 4–8.

not necessarily result in reduced rates of return for all concerned. Market-share rivalry, with profit constraints, therefore, was extended to an increased number of city-pairs. Because regulation does not directly restrict the scheduling and frequency of flights once route approval has been granted to a carrier, the market-share rivalry in recent years has meant decreasing load factors and lower rates of return to the Big Four trunk operators.

Finally, viewed over short periods, when particular nonprice competitive tactics prevail, the market-share rivalry has taken on forms that appear to increase costs. The episodes of the final stretches of the DC-7 and Lockheed Super Constellation series and the product differentiation that leads to luxury food, beverage, and entertainment services are examples. Nonetheless, aircraft equipment choices and, presumably, the methods of providing services are strongly influenced by cost considerations. In the period 1947–65, the number of particular models of planes purchased in any year by trunk-line carriers is significantly associated with the relation of the seat-mile operating costs of the particular model to those of other models that had similar performance characteristics and were available at the same time.[56] And regardless of their other characteristics, aircraft such as the Douglas DC-5 (1940) and DC-8 (1946),[57] the Curtiss-Wright Transport (1940), Lockheed's Saturn (1946), and Republic's Rainbow (1946) had potentially high operating costs and failed to enter domestic fleets. Thus the evidence suggests that operating costs are not fully subordinate to speed, range, size, passenger comfort, and other factors in aircraft investment choices.

Summary and Speculations

From the beginning commercial air transport has undergone rapid technological change. But except for changes relating to safety and arising from the 1934 legislation separating operation from manufacturing, available evidence suggests that regulation has not had a major effect on

56. These findings are preliminary at this writing. Indications are that the share of new deliveries of a particular model in a particular year decreases by more than 1 percent for each percentage point of increase in relative operating costs. Final results will appear in the author's *Technology and Market Structure*.

57. This is not the DC-8 jet aircraft of the late 1950s, but rather a piston-engined, small commercial aircraft designed for the feeder-line and short-haul market.

technological change. The important changes originated outside the industry.

At the same time, regulation has been affected by technological change, but with a considerable time lag. Federal regulatory agencies were given the power to control market structure directly at least a decade after the need became apparent. Indeed, regulation came so late that the reversal in policy—from entry foreclosure to the use of competitive criteria in handling applications for new routes—was again perhaps a decade late.

If, as has been suggested here, the federal regulation of commercial air transport has reflected little except a mediating influence among the principal carriers, including the mediation of differences caused by varying expectations of long-term and short-term gains, these lags in regulation are not unusual. In the history of many industries visible events that ought in theory to affect market behavior in particular ways often occur long before the ultimate adaptation to these events. "Too little, too late" applies to the behavior of the air transport industry in the United States as well as to the political reactions of nations.

Not at all inconsistent with this view of technology and regulation are Gellman's findings that the "air carriers were profoundly and wrong-headedly influenced in their choices of flight equipment by the amount and character of regulation imposed upon them by the Civil Aeronautics Board."[58] The point to note is that regulation by the CAB was unlikely to result in a major protest from the carriers; in a broad sense, the regulation was of the carriers and for the carriers, if not by the carriers. The wrong-headedness was really their own.

It is not clear what regulatory alternative would have been socially preferable. Military and other types of research and development were continually making new types of aircraft possible. These new types did not always permit improved performance *and* lower seat-mile costs. They did not always permit lower seating densities and new varieties of passenger services that attract at least some air travelers *and* reduce seat-mile costs. They did permit combinations useful in nonprice rivalry. What is the optimal trade-off between higher speed and seat-mile costs? Between longer range and seat-mile costs? Between low-density seating

58. Gellman, "Effect of Regulation," p. 85. The wrong-headedness relates to purchase of aircraft such as the DC-7C and the Lockheed L-1049C and the later Constellations. These planes increased operating costs but were bought because of nonprice rivalry.

and seat-mile costs? When passengers do not have a free choice among the full range of possibilities, these questions cannot be answered. The carriers, as might be expected, persuaded aircraft manufacturers to delve into the possible technical frontiers and to propose new aircraft that would perform to the carriers' advantage. It is hard to say whether these were also in the public interest.

One can say with considerable certainty that technical progress in commercial aviation in the years ahead will be different and, by the criteria used above, slower than in the past. Research and development for commercial aviation can go in several directions. Aviation research into larger or faster or longer-range fixed-wing aircraft probably does not have the military justification it has had in years past. In particular, some of the highest-ranking military counselors have seen no need to develop new bombers, including supersonic varieties.

Without military development in advance of commercial development, the costs to society of achieving higher-performance, fixed-wing commercial aircraft will be tremendous. Military technology has advanced to the point where subsonic jumbo jets have become commercially feasible. It has probably advanced so far that the additional costs necessary to bring forth a U.S. version of a supersonic transport in the Mach 2 to Mach 3 performance range will be undertaken before foreign versions of such aircraft wholly dominate the market. But with the record of regulatory lag in the federal government, significant developments beyond this appear unlikely.

Table 5-9. Airlines Entering Scheduled Domestic Service, 1914–38[a]

Year	Airline	Initial routes
1914– 25	Aero Limited (1919)	New York-Atlantic City; Miami-Nassau
	Aeromarine West Indies Company (1919)	Key West-Havana; New York-Atlantic City (1921); Detroit-Cleveland (1921); Miami-Nassau (1922)
	Sid Chaplain Aircraft Company (1919)	Los Angeles-Catalina
	Pacific Marine Airways (1922)	San Pedro-Catalina
	Ryan Airlines (1925)	Los Angeles-San Diego
	St. Petersburg-Tampa Airboat Line (1914)	St. Petersburg-Tampa
1926	Colonial Air Transport	New York-Boston
	Colorado Airways	Cheyenne-Pueblo
	Florida Airways Corporation	Atlanta-Miami

Table 5-9 *(continued)*

Year	Airline	Initial routes
	Ford Motor Company	Detroit-Chicago-Cleveland
	National Air Transport	Chicago-Dallas
	Northwest Airways	Chicago-St. Paul
	Pacific Air Transport	Seattle-San Francisco-Los Angeles
	Philadelphia Rapid Transit Service	Philadelphia-Washington
	Robertson Aircraft Corporation	Chicago-St. Louis
	Stout Air Services	New York-Boston
	Varney Air Lines	Elko-Pasco
	Western Air Express	Salt Lake City-Los Angeles
1927	Clifford Ball[b]	Cleveland-Pittsburgh
	Boeing Air Transport[c]	Chicago-San Francisco
	Colonial Western Airways	Cleveland-Albany
	Embry-Riddle	Cincinnati-Chicago
	National Air Transport	Chicago-New York
	St. Tammany-Gulf Coast Airway	Atlanta-New Orleans
	Standard Air Lines	Los Angeles-El Paso
1928	Braniff Air Lines	Tulsa-Oklahoma City
	Canadian Colonial Airway	Montreal-New York
	Capitol Airway	Chicago-Indianapolis
	Central[d]	Tulsa-Wichita
	Commercial	Seattle-Vancouver
	Continental Air Lines[e]	Louisville-Cleveland
	Interstate Airlines	Atlanta-Chicago
	Jefferson	Rochester-Minneapolis
	Maddux Air Lines	Los Angeles-San Francisco
	Midwest	Waterloo-Des Moines
	Mutual	Los Angeles-Oakland
	National Parks Airways	Salt Lake City-Great Falls
	Northern	n.a.
	Pitcairn Aviation[f]	New York-Atlanta
	Rankin	Portland-Yakima
	Southern Air Transport	Atlanta-New Orleans-Houston
	Spokane	n.a.
	Texas Air Transport	Galveston-Dallas-San Antonio
	Thompson Aeronautical Corporation	Chicago-Bay City-Detroit-Cleveland
	United States Air Transport[g]	Washington-New York
	Universal Aviation Corporation[h]	Chicago-Cleveland
	West Coast Air Transport	San Francisco-Portland-Seattle
	Wichita	Kansas City-Wichita
1929	Atlantic Coast Airways	New York-Atlantic City
	Brower Air Service	Wichita-Omaha
	Continental Air Express[e]	Los Angeles-Alameda
	Curtiss-Wright Flying Service	Chicago airport-Grant Park

162

Table 5-9 *(continued)*

Year	Airline	Initial routes
	Delta Air Service	Dallas-Shreveport-Monroe-Jackson-Meridian-Birmingham
	Eastern Air Transport	New York-Atlanta
	Gorst Air Transport	Seattle-Bremerton
	Kohler Aviation Corporation	Grand Rapids-Milwaukee
	Mamer Air Transport	Portland-Spokane
	Mason and Dixon Air Lines	Cincinnati-Detroit
	Mid-Continent Air Express[i]	Denver-El Paso-Kansas City
	Middle States Air Lines	Akron-Detroit-Pittsburgh
	Nevada Airlines	Los Angeles-Reno-Las Vegas
	New Orleans Air Line	New Orleans-Pilottown
	Pickwick Airways	San Diego-Los Angeles
	Pittsburgh Airways	Pittsburgh-Newark
	Rapid Air Lines	Rapid City-Huron
	Seagull Airlines	Salt Lake City-Ely
	Southwest Air Fast Express	Dallas-Tulsa-St. Louis-Sweetwater-Kansas City
	Transcontinental Air Transport[j]	Los Angeles-Columbus
	United States Airways	Kansas City-Denver
	Wedell-Williams Air Service	New Orleans-Shreveport-St. Louis-Grand Isle
	Yellow Cab Airways	Kansas City-Twin Cities
1930	Air Ferries	San Francisco-Alameda-Vallejo
	Bowen	Houston-Fort Worth-Tulsa; Oklahoma City-Dallas
	Clarksburg Airways	Charleston, W. Va.-Pittsburgh
	Cromwell Airlines	San Angelo-Dallas-San Antonio
	Dixie Flying Service	Greensboro-Washington
	Eagle Air Lines	Kansas City-Des Moines
	Kalispell Airways	Kalispell-Butte
	Main Flying Service	Cincinnati-Pittsburgh
	Frank Martz Coach Company	Wilkes Barre-New York
	Michigan Air Express	Grand Rapids-Harbor Springs
	New England and Western Air Transport Company	Albany-Boston
	New York Airways	Atlantic City-New York
	New York, Philadelphia, and Washington Airways[k]	New York-Washington
	Sky View Flying Service	Pittsburgh-Niagara Falls
	Trans-American Air Lines	Pontiac-Muskegon
	Transcontinental and Western Air	Los Angeles-Kansas City-Columbus
	United Air Lines[l]	Various routes of combined lines
1931	American Airways[m]	Various routes of combined lines
	Border Air Lines	Sheridan-Great Falls

163

Table 5-9 *(continued)*

Year	Airline	Initial routes
	Century Air Lines	St. Louis-Chicago-Toledo-Cleveland-Detroit
	Century Pacific Airlines	Los Angeles-San Francisco-San Diego-Phoenix
	Chicago-Detroit Airways	Chicago-Detroit
	Continental Airways[e]	Washington-Chicago
	Metropolitan Air Ferry Service	North Beach-Newark-Brooklyn
	Oklahoma-Texas Air Line	Wichita Falls-Ponca City
	Reed Air Lines	Wichita Falls-Oklahoma City-Ponca City
	Richmond Air Transport	Richmond-Washington
	Trump Airways	Little Rock-Tulsa
	Tuxhorn Flying School	Springfield-Kansas City
	Varney Air Service[n]	Los Angeles-San Francisco
	Wilmington-Catalina Air Lines	Wilmington-Avalon
	Wyoming-Montana Air Lines[o]	Denver-Billings
1932	Champlain Air Transport	Plattsburgh-Burlington
	Coast Airways	Los Angeles-Santa Barbara
	Commuters Air Service	Hartford-Springfield
	Gilpin Air Lines	San Diego-Los Angeles
	Hanford Tri-State Airlines	Sioux City-St. Paul
	Hunter Airways	Memphis-Tulsa
	Inter-City Air Lines	Springfield-Boston
	Maine Air Transport	Rockland-Stonington
	Portland Airways	Portland-Walla Walla
1933	Cardiff and Peacock Air Lines	Los Angeles-Bakersfield-Fresno-San Francisco
	G and G Airlines	San Diego-Los Angeles
	Licon Airways	Islip-New Haven
	National Airways, Inc.[p]	Boston-Bangor
	Ozark Airways	Springfield-Kansas City
	Pacific Seaboard Air Lines[q]	Los Angeles-San Francisco
1934	Central Airlines	Washington-Detroit
	Chesapeake Air Ferries	Baltimore-Ocean City
	Delta Air Lines[r]	Charleston-Dallas
	Island Airlines	New Bedford-Nantucket
	Long and Harmon	Amarillo-Fort Worth
	National Air Line System	Daytona Beach-St. Petersburg
	Robertson Airplane Service Company	Houston-New Orleans
1935	Columbia Airlines	Detroit-Louisville
	Consolidated Airlines	Alameda-Sacramento
	Land o' Lakes Airline	Detroit-St. Ignace

Table 5-9 *(continued)*

Year	Airline	Initial routes
	Watertown Airways	St. Paul-Spearfish
1936	Capital Airlines	Boise-Pocatello
	Condor Air Lines	San Francisco-Salinas
	Grand Canyon Airlines	Boulder City-Grand Canyon
	Palm Springs Airlines	Los Angeles-Palm Springs
1937	Airline Feeder System	New York-Westfield, Mass.
	Atlantic and Gulf Coast Airlines	Savannah-Mobile-Jacksonville
	Miami-Key West Airways	Miami-Key West
1938	Marquette Airlines	St. Louis-Detroit
	Mayflower Airlines	Boston-Nantucket

Sources: U.S. Department of Commerce, *Air Commerce Bulletin* (1929–38); Davies, *A History of the World's Airlines; Aviation* (Sept. 1, 1928, and May 25, 1929). Unavailability of statistics is denoted by n.a.

a. What appear to be new lines but actually represent only name changes have been eliminated as far as possible. For example, Alfred Frank Air Line, operating in 1934, is the same line as National Parks, which entered in 1928. Boston and Maine Airways, of 1937, is National Airways, formed in 1933. This subsequently became Northeast Airlines. General Airlines, 1934, is a renaming of Western Air Express, which maintained a separate existence after a 1930 merger with Transcontinental Air Transport-Maddux. Mid-Continent Airlines, 1938, is the old Hanford Tri-State Airlines of 1932.

b. This became Pennsylvania Airlines in 1930 and, after merger with Central Airlines, was eventually renamed Capital Airlines. It later merged with United Air Lines.

c. The predecessor company to United Air Lines, which is listed separately in 1930. The 1930 listing is not counted in the text discussion as a new entry.

d. Not the same as Central Airlines, a 1934 entry.

e. Not the same as the present Continental Airlines, which grew from Varney Air Service, a 1931 entry.

f. The predecessor company to Eastern Air Lines, or Eastern Air Transport, which is listed separately as a 1929 entry. In the text count of entries, the 1929 listing is not included.

g. Not the same as United States Airways, a 1929 entry.

h. Original operating company of the Aviation Corporation, which subsequently developed American Airways.

i. Not the same as the Mid-Continent Airlines of 1938.

j. This company was organized in 1928 but began service in 1929. After mergers with Maddux and Western Air Express in 1929 and 1930, it became Transcontinental and Western Air, later Trans World Airlines. The appearance of Transcontinental and Western Air in 1930 is not counted in the text as a new entry.

k. Also known as Ludington Air Lines in 1931 and 1932. The line disappeared in a merger with Eastern.

l. Formerly Boeing Air Transport. Not counted in the text as a new entry.

m. American Airlines after 1934. The 1931 appearance is not counted in the text as a new entry.

n. A reentry by Varney after the Varney Air Lines of 1926 was absorbed by United.

o. Subsequently Wyoming, and later Inland Air Lines.

p. Subsequently Boston and Maine Airways, and later Northeast Airlines.

q. Subsequently Chicago and Southern Air Lines.

r. A reentry of Delta after suspension of Delta Air Service of 1929. Not counted in the text as a new entry

CHAPTER SIX

Surface Freight Transportation

Aaron J. Gellman

THIS CHAPTER COVERS the impact of economic regulation on firms carrying freight by surface transport. It raises questions about the effect of regulation on the propensity to innovate of the regulated carriers and of the firms that supply their equipment. It then examines the effect of regulation on the market structure in surface transport and the influence this may have on innovation by the carriers, their suppliers, and their customers. Where regulation is found to have discouraged or distorted innovation, remedies are proposed.

The Scope of Freight Traffic Regulation

How pervasive is economic regulation of freight transportation, and what are the principal techniques employed? Estimates of the extent of federal regulation, by volume and by revenues of intercity freight transportation, are shown in Table 6-1.[1] Only summary information, devoid of detail, is presented. For example, the quality of transportation under regulation is neglected. The data show that producers of the various surface modes of transportation, except the railroads, do not always have to contend with regulation when making investment or pricing decisions—decisions in which innovation is usually an issue.

Carriers in the various competing and complementary modes of transportation, even when they are subject to federal regulation, are con-

1. The inclusion of state and local regulation would have a significant effect on highway carriers only.

Table 6-1. Ton-Miles and Revenues of Domestic Freight Transportation Subject to Federal Economic Regulation, by Mode, 1970

In percent

Mode	Ton-miles	Revenues
Rail	100	100
Highway	35	50
Water	10	15
Pipeline (crude oil and petroleum products)	100[a]	100[a]

Source: Author's estimates, based on data from the Interstate Commerce Commission and the U.S. Bureau of the Census, *Census of Transportation, 1963.*

a. Economic regulation of pipelines relates entirely to rates and then only in a general way, since the objective of the ICC is to assure that a fixed rate of return on invested capital is not exceeded.

strained in different ways. Table 6-2 shows some of the sources of intermodal differences in regulation. In addition, some types of regulation are quite precise and pervasive in one mode but sketchy or nonexistent in another. Because of the mixed character of regulation, comparative analysis is difficult. Most of the examples of surface transportation used in

Table 6-2. Regulation of Interstate Domestic Surface Carriers, by Type and Mode, 1970

Type of regulation	Modes			
	Rail	Highway	Water	Pipeline
Rates	Full	Partial[a]	Partial[b]	Partial[c]
Entry or exit through carrier expansion or contraction	Full	Full	Full	None
Entry through merger or acquisition, same mode	Full	Full	Full	None
Entry through merger or acquisition and other mode	Partial	None[d]	None[d]	None[d]
Overall quantity of service	None	None	None	None
Quality of service, including route	None	Full	None	None
Limitations on commodity and direction-of-traffic	None	Full	None	None

a. The degree of rate regulation is largely a function of the form of carrier organization. Common carrier rates are regulated most; agriculturally exempt and "private" carrier rates (or internal prices) are not regulated. Contract carriers fall in between.

b. This is similar to highway regulation, except that exemptions are granted for other reasons besides the agricultural commodities provision.

c. Rates are regulated on a rate-of-return basis; hence individual rates are the object of very little regulatory concern.

d. There is no explicit prohibition. Entry is subject to ultimate interpretation by the Interstate Commerce Commission and the courts.

this chapter are drawn from the rail and highway sectors, where there is greatest comparability.

Why Innovate?

What are the motives for innovation, given the objective of maximizing the firm's long-run profits? Basically, there are two polar cases. In the first, innovation can lead to a shift of the demand curve that moves the firm to a more profitable equilibrium price and output. Second, innovation can shift the average cost curve so as to increase the profitability of producing the equilibrium output. Of course, innovation can result in shifts in both the cost and demand schedules, yielding a more profitable equilibrium price and output.

Regulating Surface Transportation

Many discussions of the general effects and rationale of transport regulation may be found in books and professional journals, and this material does not need to be reviewed in detail here.[2] In brief, it may be said that, with one notable exception, economic regulation in the transport sector of the American economy reduces the managerial and operating freedom of carriers and their shippers and limits the freedom of others to enter the field. Even the exception, which grants competing rail and highway transportation firms freedom to collude in the establishment of intramodal rates and other elements of price, is per se anticompetitive and therefore fits the pattern of most transport regulation, a basic objective of which is to reduce competition. But the heart of the matter is the impact the reduction in competition in the transport sector has had on the amount and kind of innovation in transportation and in its peripheral industries, both on the buying and on the selling side.

Because there is a dearth of material specifically relating technological change and the innovation process to regulation in surface transportation, specific case studies should prove useful. Equipment histories in

2. See, for example, John R. Meyer and others, *The Economics of Competition in the Transportation Industries* (Harvard University Press, 1959), especially Chaps. 1 and 4; Emery Troxel, *Economics of Transport* (Rinehart, 1955), Chaps. 14–18, 25; D. Philip Locklin, *Economics of Transportation* (5th ed., Irwin, 1960), Chaps. 10–26, 31–34, 36–38.

several of the transport fields are of special interest, since capital investment decisions for carriers frequently cut across several regulatory lines. The underlying hypothesis is, of course, that managers of transport enterprises are profoundly influenced by the amount and character of regulation to which their industry and firm are subjected. To be sure, regulation is not the sole factor influencing decisions on investment in equipment; many other forces are also at work, including demand, capital availability, and the state of technical knowledge, among others. But regulation plays an important role in investment decisions and in the process of innovation within the transport sector.

Regulation and the Railroads

Minimum rate regulation, as applied to the railroads of the United States by the Interstate Commerce Commission (ICC), is an especially pernicious form of regulation, profoundly affecting research, development, and the innovative propensities of the carriers and their suppliers. For example, specialized innovative rail equipment will have cost characteristics associated with its use that are different from those of preexisting equipment. New equipment should produce transportation at a lower long-run average cost than old, although this may not always be the case. Especially where lower unit costs are associated with increased production, the volume necessary for profitable operation can often be generated only with lower rates. The ICC is inclined to judge proposals to reduce rates largely by the extent to which they are compensatory to the carrier. This implies that the commission has a yardstick for determining the cost with which the rate is to be compared and judged. And so it does. The best known of the several formulas is Rail Form A, which purports to measure out-of-pocket costs associated with any actual or prospective move. In fact, Form A is an average cost concept used in many instances where a marginal cost determination would be more appropriate and would lead to better resource allocation.

In general, Rail Form A (as well as other commission formulas) relies heavily on historic averages in apportioning joint costs. When the averaging concept is applied to a situation in which a carrier would rather use an improved piece of equipment with lower-than-average cost, the higher cost computed on the basis of historic averages may prevent rates from being lowered enough to attract the volume of traffic needed to justify the development and acquisition of the innovative equipment. Unfortunately, Rail Form A has killed off several railroad car ideas because costing on an average basis did not permit rates that could produce the necessary volume of traffic. In an industry that is in need of technological improvement and whose performance influences so many other enterprises and industries, this situation is frustrating to shippers, carriers, and the public.

The application of rigid cost formulas has still another inherent fault. New equipment—even when embodying substantial innovation—does not automatically produce either lower average cost or lower marginal cost. Yet on occasion, the use of the cost formulas that are forced on the railroads causes them to invest in equipment that, although new, may be less efficient than either older equipment or an entirely different type of new equipment. The latter may not be considered because of the need to show that it would improve efficiency in terms of the "approved" cost formulas.

The Piggyback Decision

A most dramatic illustration of this point is found in the decision to invest in the piggyback rail car. In the late 1950s trailer-on-flatcar services were first widely offered by U.S. railroads. Previously, piggybacking had been tried on a very limited basis using conventional railroad flatcars. Then it became necessary for the railroads to acquire rail vehicles explicitly for piggyback and related services. While many factors influenced the piggyback decision, none was as powerful as the rate-regulatory philosophy of the ICC as applied to the railroads.

In 1958, a conventional 45- to 50-foot flatcar, suitable for handling a single 40-foot trailer and with a stanchion to support the fifth wheel of the trailer, could be acquired for about $9,000. At the same time, at least two car builders began producing an 85-foot car equipped to accommodate two 40-foot highway trailers. This double-length car was offered at about $15,000. In terms of "static" capacity the new, double-length flatcars represented a saving of about 17 percent in capital cost. Far more significant was the impact of the introduction of this new equipment on Rail Form A costs—the principal reason for the railroads' enthusiastic reception of the double-length flatcar. That is, since the costs per car dictated by the formula could be spread over more traffic units (ton-miles) with the larger cars, Form A costs were lower and therefore supported lower rates than would otherwise be sustainable.

Some railroad traffic executives understood that the demand for piggy-back transportation was somewhat price elastic. To build the volume necessary to make piggybacking efficient, the rates needed to be lower than those that could be offered users of conventional single-length flat-cars. The railroads knew that Rail Form A costs based on single-length flatcars would be so high that the ICC would not permit rates low enough to attract the necessary traffic. In any case, once a few key rail-roads published rates based on these double-length cars, the rest of the industry, unable to offer lower or even equal rates for the single-length cars, was forced to adopt the long piggyback car.

Was Rail Form A correct in showing that the cost of transportation with double-length flatcar equipment would be materially lower than the cost that could be achieved with the shorter cars? For the total U.S. rail-road system, the answer is clearly no, although for *some* carriers costs have certainly been lower. Enough railroads have experienced higher costs to support the view that average unit costs associated with piggy-back transportation are higher than they would—and should—have been. In fact they are higher in many instances than Rail Form A in-dicates.

For several reasons costs have been higher with the use of the double-length flatcars for piggyback and related traffic. First, the exceptional car length and the location of the trucks well in from the end of the car re-sult in instability under many of the conditions of grade or curvature, or both, and high draw-bar pull encountered in the United States. Derailments resulting from this instability have been frequent,[3] and the associated costs are obviously significant. Second, many railroad yards and terminals—even quite modern ones—were not designed for handling cars of this length.[4] In particular, many partially automated operations must revert to old-fashioned, manual techniques when long cars are used. Again, higher-cost operations are the unavoidable result. Third, the savings in capital cost per unit of static capacity associated with the ac-

3. The trucks are so located to permit movement of the cars around the tight curves often found on older railway systems and in mountainous areas. See Association of American Railroads, Research and Test Department Report ER-17, "Running Tests of a Flat Car Trailer Carrier and a Three-Level Auto Carrier on the Burlington Railroad" (Chicago, 1965); Lewis K. Sillcox, "Track and Truck," in *The Dynamics and Economics of Railway Track Systems* (Chicago: Railway Systems and Management Association, 1970).

4. R. J. Dullard, "TOFC Figures—Are They for Real?" in *TOFC and Con-tainerization* (Chicago: Railway Systems and Management Association, July 1963), pp. 24–27.

quisition of double-length flatcars have been largely frittered away because of the indivisibility of the long car for use where only one trailer (or container) is to be moved between two points. A large proportion (some say over 20 percent) of the reported revenue car-miles[5] of piggyback equipment is produced with half the length of the car unused.[6] If there is substantial half-empty revenue car mileage, the capital cost saving per unit of static capacity becomes strictly illusory, and perhaps the capital–output ratio for the longer cars is less favorable than for the shorter units.

It is reasonable to ask whether anyone in the railroad industry at the time the piggyback equipment decision was being made foresaw the needlessly high costs that would come from widespread adoption of the longer cars. A few—but only a few—did, and most of them were not in a position to influence the decision, either because they were not in top positions or because their railroads were not strategically situated in the U.S. system, either geographically or in volume of originated traffic.

In the piggyback decision, the primary cause of distortion is the use of the car-mile in many of the Rail Form A calculations, as well as the use of historical *average* car-mile costs. Costs of the past are used in the formula, even when equipment and techniques that differ greatly from the average are being analyzed. Use of the car-mile as an allocating base in connection with double-length piggyback equipment was criticized in 1960:

Take that popular allocating base, the car mile. This is a fine old method that railroad costers use to allocate all sorts of costs, but just what is a car mile? In the old days when car lengths were pretty much uniform, it wasn't too hard to tell what a car mile was. But is an 80-foot flatcar mile the same as a 40-foot flatcar mile? Can you bolt two 40-foot flatcars together, and thereby reduce two car mile costs into one car mile cost? Or would it be more realistic to do things on the basis of car foot miles? Is it possible to run 200-car trains if the cars are 80 feet long? That is 16,000 feet of train, or three miles. Can such a train be properly yarded?

To the extent that car mile costs are based on average train lengths, would

5. A revenue car-mile is recorded and reported to the ICC whenever a car under load is moved one mile, regardless of what percentage of the car's load limit is used. It is a poor measure of railroad capital utilization but is used by the ICC and others. Of the mass of statistics collected by the ICC, none relate revenue car-mile data to load factor.

6. "The Growing of a Half-Billion-Dollar Industry," *Modern Railroads,* Vol. 24 (November 1969).

it not be proper to drastically increase car mile costs when the cars get longer than average? Let's put it bluntly: are we not undercosting the 80- and 85-foot cars that everybody is using for piggy-back?[7]

Rail Form A helped to mislead the railroads in another way about the piggyback method of transport. It is now frustratingly clear that the cost of producing rail transportation using a trailer-on-flatcar technique is considerably higher than the cost using the container-on-flatcar technique.[8] Yet the Rail Form A costs, on which the rates must be based, are the same where identical rail equipment is used and the same traffic and routing are assumed. Rail Form A fails to take into account that by virtue of the laws of aerodynamics fuel costs are substantially greater where trailers are being hauled. At 60 miles an hour, with trains of identical weight and with the same double-length flatcars, fuel consumption of trains with trailers is about three times that of trains with containers flush on the deck.[9] At the same time, the motive power required to haul the trainload of containers is substantially less. The effect of Rail Form A's failure in this regard is partially evident when the railroads publish identical rates for container and for piggyback traffic, even though the costs differ greatly. Of course, where cost savings are not reflected in rates, there is economic discrimination per se. It is built into the rate-making system only partially by Rail Form A; but as railroaders themselves become aware of the need to identify these cost differences before tariffs are published and freight is moved, Rail Form A becomes the principal barrier to nondiscriminatory pricing.

Rail Form A distorts costs in still another way. Since the formula uses broad average costs based on the use of conventional equipment, its application to wholly new equipment and situations usually leads to results that are unfortunate for carrier and shipper alike. Many of the events that determine railroaders' decisions about technology, marketing, and rate-making are discrete, random occurrences, such as a major innovation in a competing mode or by a user of transportation services. Reliance on

7. John W. Ingram, "The Need for Improving Costing Concepts and Techniques," in *Economic Costing of Railroad Operations* (Chicago: Railway Systems and Management Association, 1960), p. 11.

8. "The Box Worth Billions," *Forbes* (April 1, 1968), pp. 30–31; "Containerization: Always a Bridesmaid," *Railway Age* (Dec. 12, 1966), pp. 20–21.

9. W. L. Paul, "Containerization: Concepts and Research on the Santa Fe," *Ninth Annual Meeting, Supplement*, Transportation Research Forum (Chicago, 1968), pp. 165–66.

slowly adjusting historical formulas to determine costs slows the rate at which the railroads adapt to these events. Lower costs are attainable long before carriers can take full advantage of them because the lower costs are mixed with others in the averages that the ICC uses in its formulas.

Rail Form A, because of its cost formulas, is clearly a significant barrier to innovation. The propensity to innovate on the part of both carriers and suppliers is seriously curbed by the very existence of such formulas. Moreover, because economic regulation based on cost formulas like those of Rail Form A is very rigid, it can be shown that the kind of technological change that would have a salutary effect on transport system operations is discouraged while new techniques that offer questionable benefits are fostered. In short, regulation by formula distorts management's innovative drives as much as it frustrates them. For example, if as much energy had been poured into development of the optimum piggyback vehicle, from the point of view of traffic and operations, as was devoted to finding the vehicle that would yield the lowest Rail Form A costs, it is safe to assume that the cost—and service—aspects of railway piggyback operations in the United States would be far more attractive today than they are.

The Regulatory Review Delay

Another powerful influence on transport innovation is the delay in innovating caused by regulatory review, which frequently causes significant time to elapse between the conception and the exploitation of an innovation. Since one of the principal motives for innovating is to foster greater demand for one's product or service, the timing of an innovation can drastically affect the profitability of any given investment in innovation. If a firm innovates successfully, it reduces its costs or increases the demand for its product or service or both; it may be able to capture the benefits of innovation only if it can get a time jump on its competitors. Any external force that causes the innovator to announce his intentions beforehand will reduce the time in which he can realize the highest possible level of profits from the innovation. While advance news of a prospective innovation does not eliminate its profitability, especially not in the case of a cost-reducing innovation, the profitability may be reduced sufficiently to make the innovation unattractive when risks and alternative uses for the capital are considered. If the risk of an innovation is increased, future profits are discounted at a higher rate, driving down the present discounted value of the profits of innovation. The regulatory scheme under which the railroads and, to a lesser extent, other regulated surface carriers operate has precisely this effect. The regulatory delay impinges on the process of innovation for regulated carriers largely by linking the application of new technology with the need to publish new and often conceptually different tariffs governing the use of such technology. Delayed introduction or expansion of service with larger railroad tank and covered hopper equipment, automobile-rack cars, and other special-purpose units are cases in point.

Because the process of securing approval of rates is usually a long one, especially where there is a break with tradition in either the form or the structure of tariff publication, the inevitable result is to dampen the enthusiasm of carrier management for technological innovations that, in order to be profitably exploited, must be subjected to the regulatory process. For the railroads perhaps the most difficult case occurs when innovative rolling stock creates substantially lower service costs than conventional equipment.[10] As was indicated above, the problem arises principally because the innovation reduces costs to a point below what is considered to be compensatory by the ICC, using the Rail Form A historic-cost formulation.

The celebrated "Big John" grain rate case of the Southern Railway is a classic example.[11] In 1961 the Southern published a highly innovative rate structure covering shipments of relatively large quantities of grain into points in the southeastern United States served by the Southern from crossings of the Ohio and Mississippi rivers. These rates were to become effective in August 1961. Some indication of the revolutionary nature of these changes is given by the fact that the *average* reduction brought rates about 60 percent below the previous rates for such traffic shipped entirely by rail. To qualify for the new rates, the shipments had to be made in unconventional, lightweight, covered hopper railway cars and

10. For example, costs may be reduced by using cars that have dramatically lower empty weight (aluminum instead of steel for the car body), that require less maintenance (roller bearings), and that can be loaded and unloaded more quickly and with less labor (pressure differential covered hopper equipment). Most difficult and frustrating are the cases where the savings are in components of the system that are not reflected in the cost formulations.

11. "The Southern Railway's 'Big John' Grain Rates," talk by John W. Ingram at the Transportation Research Forum, New York, Feb. 2, 1965.

had to be in substantially larger quantities than were required for the higher boxcarload rate. To receive the new rate, the shipper also had to forgo transit privileges,[12] but the Southern had determined, before publication of the rate, that less than half of the railroad's previous grain traffic took advantage of transit privileges. Moreover, the Southern study showed that barge traffic of grain moving into the Southern's territory was increasing far more rapidly than parallel rail shipments of grain. The railroad had good reason to believe that the movements of grain by truckers were also increasing rapidly. Southern's response was to devise a scheme that would use the multiple car movement of 100-ton aluminum covered hopper (Big John) cars and sharply reduce "free time" for loading and unloading.[13]

As a result of objections raised by numerous shipping and receiving firms and, most significantly, by the barge lines,[14] the ICC set aside the Southern's Big John rates.[15] On the basis of a hearing beginning in January 1962 and concluding in August 1962, a division of the ICC approved the rates.[16] After reviewing the decision, the full commission reopened the proceeding and on July 1, 1963, reversed the endorsement of the Big John rate scale.[17] After a long, expensive, frustrating series of legal maneuvers and further hearings, including a Supreme Court order for the commission to reconsider its earlier decision,[18] the ICC, on September 10, 1965, substantially upheld the earlier approval of the Big John tariff.[19]

More than four years passed between the announcement by Southern of its intention to introduce the innovative Big John aluminum hopper

12. Transit privileges permit shipments to stop off in transit for varying periods of time for storage or further processing or manufacture, or both, before moving on to market.

13. Free time is the period allowed shippers following placement of a car for loading or unloading before demurrage charges are assessed. Under the Southern Big John tariff, free time was reduced from forty-eight hours to twenty-four hours.

14. "Petition of Protestants for Reconsideration of the Report and Order of Division 2, before the ICC, Investigation and Suspension Docket 7656, 'Grain in Multiple-Car Shipments—River Crossing to the South,'" submitted by the Waterways Freight Bureau and its members, and also to the American Waterway Operators, Inc., and its members, and others, Feb. 25, 1963.

15. ICC Investigation and Suspension Docket 7656, of rates originally scheduled to become effective Aug. 10, 1961.

16. 318 ICC 641, Jan. 21, 1963.

17. 321 ICC 582, July 1, 1963.

18. 372 U.S. 658, April 15, 1963; 379 U.S. 642, Jan. 18, 1965.

19. 325 ICC 752, Aug. 30, 1965.

cars for moving grain and the decision to permit the railroad to carry out its original plans and intentions. The Southern had placed some Big John equipment in service hauling grain during this period, and had even applied the sharply reduced rates because the Interstate Commerce Act provides that after seven months of suspension a carrier may make a rate effective if it has not been set aside. But full exploitation of the lower rates called for substantial capital investment by shippers and receivers of grain. Nothing approaching full profitability could be derived by the Southern from the combined innovations of the Big John grain cars and the rates needed to make them work in the marketplace. These rates, as noted, would have been dramatically lower than those previously available for all-rail movement and, given the shape of the demand curve facing the Southern, would have led to a dramatic increase in the railroad's movement of grain.

The Big John rates might never have been approved had it not been for a fortuitous circumstance. Late in 1962 the Commodity Credit Corporation (CCC) undertook to move large quantities of grain to meet a shortage in the southeastern United States. The Southern offered Big John service under the rates and conditions originally proposed in 1961. The CCC accepted the Southern's offer on a Section 22 basis.[20] This gave the railroad an opportunity to show that it could produce this type of transportation at far lower costs, a fact that was being challenged before—if not by—the ICC.

As if these regulation-imposed barriers to innovation were not enough, during 1960 and 1961 the Southern acquired some $7 million worth of Big John covered hopper units for use with the new, reduced rates for grain. By the time the Big John rate structure was finally approved and the railroad, the shippers, and the receivers could safely make the additional capital investments needed to take full advantage of the revolutionary rate structure, the Southern had invested over $30 million in Big John cars alone. The extent to which this equipment was underutilized while the regulatory review was proceeding cannot be measured, but obviously the profitability of the investment was less than the Southern could have realized had the new rates been approved in August 1961. In

20. Section 22 of the Interstate Commerce Act permits carriers to negotiate and publish different rates for government agencies than are available for the same service to other shippers. Section 22 rates are not subject to ICC review and the consequent regulatory delay.

addition to the investment in equipment, the Southern invested in fixed plant in support of Big John operations, and the legal and administrative costs of prosecuting the case ran to millions of dollars.[21]

The Southern Railway's Big John case is one of the few instances in American railroad history in which management has had the determination to see the regulatory process through. The increases in traffic volume under the Big John tariff, together with the reduced cost to the railroad of producing the transportation, tend to bear out the Southern's original contention about the demand for and profitability of this service. Innovations of comparable or even greater importance have been blocked by the barriers to innovation erected by the regulatory process, especially the regulatory delay.

Constraints on Highway Transport

As Table 6-1 indicates, a significant portion of intercity freight transportation produced in the highway mode is subject to economic regulation. All of it is subject to safety and load limit regulations imposed by federal, state, and local agencies. While regulations limiting the weight and size of vehicles have important economic implications, these will not be analyzed in this chapter.

In part because highway carriers in maximizing profits have not found necessary or desirable the publication of rates lower than those of the railroads, their principal competitors, the question of minimum rate regulation of highway carriers has been largely academic. One result is that in no important instances have highway carriers needed explicit rate approval from the ICC to be able to introduce a technological innovation with the impact of, say, the Big John covered hopper grain cars or the double-length railway equipment. Unlike the railroads, the constraint on the highway carriers relates to the physical limitations of the roadway. Indeed, in a recent paper dealing explicitly with highway equipment innovation, Robert S. Reebie notes that what the highway carrier needs most is cost-saving innovation.[22] The Reebie paper does not suggest that

21. "Southern Victory Swamps 'Invincible' Bureaucracy," editorial by John F. Yarbrough in the *Georgia Poultry Times* (Sept. 22, 1965); "Potential for Growth in Rail Transportation of Grain," talk by Robert C. Haldeman at the Minneapolis Grain Exchange Marketing Seminar, Sept. 7, 1966.

22. "Highway Equipment Innovation for Improved Public Service and Carrier Profits," in *Papers—Eighth Annual Meeting: Man and Transportation*, Transportation Research Forum (Richard B. Cross Co., 1967), pp. 181–94.

stress is being laid on innovations that might increase the demand for highway transportation except as weight and dimensional limitations are independently liberalized. Because of the constraints on operating practices and the associated limitations on profitability inherent in the weight and dimensional restrictions, Reebie suggests a number of ways of developing equipment with a lower empty weight and of using the currently available cubic envelope to greater advantage. The only other "innovation" suggested, which would improve highway carrier efficiency but which is constrained by current regulatory practices, relates to the legal prohibition against carriers obtaining backhauls in many cases. This locks them into the wasteful practice of returning with empty equipment and is probably the most severe regulatory constraint on innovation in highway transport. General commodity carriers could often obtain a return load if they could use large collapsible rubber tanks to haul either dry or liquid bulk cargoes as well as dry freight. Where they cannot do so, it is because their certificates of public convenience and necessity limit the commodity they may carry or the direction in which they may move it. For example, highway carriers of new automobiles frequently have operating authority from one or more points to other points but no "between" authority. In most instances they can haul no other commodity, which removes any incentive to develop trailers that will accommodate vehicles as well other commodities on return trips or at times when automobile production has declined.

The ICC seems never to have considered the inevitable inefficiencies in one-way and/or limited-commodity certificates. Instead, the commission defends its policies in terms of existing methods and cost functions and ignores the possibilities for economies inherent in more advanced technology. Perhaps if collapsible rubber tanks capable of hauling a great variety of liquid or dry bulk freight had been in existence when certificates were being awarded to dry freight carriers, the commission would have permitted such tanks, when filled, to be classified as general commodities under the carriers' certificate provisions. But the ICC, in highway as in railroad transport, has tended to ignore both the technological realities and the technological possibilities, while establishing policies that have become virtually permanent.[23]

23. The situation is made even worse by the existence of several classes of highway carriers besides common carriers, each with different restrictions. The common carrier undertakes to carry all freight over regularly scheduled routes at nondiscriminatory rates. The contract carrier transports certain goods over the

Within the size and weight constraints, truckers have constantly sought to combine the advantages of relatively large pieces of highway equipment that produce substantial increments of transportation with the efficiency in handling and the market attractiveness of units smaller than those that are operated by their principal competitors, the railroads. While containerization has helped truckers to a limited extent in this respect, the recent liberalization in many states of restrictions on the size and weight of vehicles to permit operation of double-bottom and triple-bottom rigs has been far more important. The technology at work relates to more powerful tractors, better highway construction, and improved control of trailing vehicles (especially when braking). Perhaps even more important is a better understanding by state highway authorities of just what size and weight restrictions are needed. With the relaxation of these restrictions, substantial new technological innovation will probably be introduced to attract new commodities and new shippers to highway carriers, as well as to increase reliance on the highway mode by shippers who now split their patronage between the highway carriers and other modes.

Another desirable regulatory action would be the removal of the barriers to intermodal cooperation. This action would stimulate technological innovation in both transport vehicles (including containers) and container-handling or trailer-handling equipment. Indeed, widespread publication of tariffs encouraging the intermodal movement of vehicles or containers, regardless of whether they are owned by carriers, shippers, or leasing companies, would invite substantial technological innovation sponsored by common and private carriers and by shippers and lessors. For example, many private carriage operations now carried out entirely by road might more efficiently be produced intermodally with proper regulatory encouragement and further technological development in trailers and containers. Of course, many of the negative or distorting effects on innovation in intermodal transportation that stem from regulation could be alleviated by creating a number of competitive, horizontally integrated transportation companies that either produce transportation directly or

route and at the price designated in a contract or agreement. A "private" carrier handles the owner's goods in furtherance of a commercial enterprise. Agricultural exemption applies to motor vehicles used for transportation of farm products, or by a farmer or agricultural cooperative association, all of which are exempt from the regulatory provisions of the Motor Carrier Act of 1935. See Julius H. Parmelee, *The Modern Railway* (Longmans, Green, 1940), pp. 490, 547, 553.

contract for the services of independent carriers in any and all modes of transportation. W. J. Stenason of the Canadian Pacific Railway has said:

I would suggest, for example, that the earlier large-scale introduction of piggyback service in Canada, as compared with the experience in the United States, can be substantially explained by the lack of restrictions in Canada which limited multi-modal ownership, compared with very onerous multi-modal restrictions in the United States. Integrated transportation firms in Canada were able to capitalize on the obvious economies of integrated rail-highway services. Restrictions on multi-modal ownership in the transportation market in the United States led to a considerable delay in the introduction of these economies.

. . . . .

As a practical matter, the unavoidable diversity of interests which exists where a co-ordinated multi-modal approach is necessary makes it most difficult to achieve a truly low-cost transportation mix unless there is common ownership.[24]

Everything considered, technological innovation in highway transportation is less severely constrained and distorted by regulation than it is in railroad transportation. Still, any interference with the innovative process should be eliminated if the public benefits are not greater than the costs. The burden of proof should be on the regulators to show that the value of the regulation exceeds its costs. The elimination of backhaul restrictions seems long overdue, and most limited-commodity certificates should be broadened. This alone would provide a substantial impetus to technological innovation and improved efficiency in highway transportation.

Inland Water Transport and Regulation

For carriers operating on the inland waterways of the United States technological innovation has been somewhat limited in recent years, partly because some of the regulations that have been applied were designed for an earlier technological environment. An important example is the provision of the Interstate Commerce Act that exempts from economic regulation movements of bulk commodities where "not more than three such commodities" are carried in a single tow.[25] In recent years, towboats of up to 8,000 horsepower have become available, yet common

24. "Multi-Modal Ownership in Transportation," *Papers—Eighth Annual Meeting,* Transportation Research Forum, pp. 470–71.
25. 54 Stat. 931.

carrier barge operators have little opportunity to use this modern, efficient equipment since they cannot compete with unregulated and private carriers by hauling more than three different commodities at one time.

Other government-imposed limits on technological development in inland waterway transportation relate to the size and draft of vessels, which are in turn largely determined by controlling depths and lock sizes, and to the subsidy to water carriers inherent in the federal provision and maintenance of the waterways. The implications for technological development of these governmental policies are not considered further here.

The Pipelines and Rate Regulation

Pipeline regulation is especially interesting for purposes of this study because it is relatively simple and the regulatory mechanism is generally thought to be irrelevant to the technology involved. Clearly, the regulators do not feel that they constrain or distort innovation in the pipeline field. Is this actually the case?

Pipeline regulation, when applied, has generally focused primarily on rates and then only as a means for ensuring that earnings do not exceed an established rate of return on invested capital. The ICC places a valuation on pipeline properties and investment, including working capital, and monitors revenues and costs so that the rate of return does not exceed 8 percent for crude oil lines and 10 percent for petroleum products lines after taxes.[26]

Common carrier pipeline ownership is vested in the principal producers and marketers of crude oil and petroleum products. The propriety of this arrangement has frequently been questioned, but as yet no regulatory action has resulted. Similarly, antitrust action has been confined to studies by the Justice Department and hearings and threats by Congress —most recently by the House Antitrust Subcommittee. No specific action has been taken recently to extend pipeline regulation or to restrict further the freedom of pipeline owners and operators.

There are several indications that even the relatively simple regulation applied to pipelines discourages innovation. To support this, some specific examples, most of which are drawn from Arthur M. Johnson's study

26. See George S. Wolbert, Jr., *American Pipe Lines: Their Industrial Structure, Economic Status, and Legal Implications* (University of Oklahoma Press, 1952), p. 135.

of petroleum pipelines, may be cited.[27] To begin with, the valuation procedure critically affects the impact of regulation on management's propensity to innovate. For example, if the rate base—that is, the capital investment on which the rate of return is computed—includes carrier property either completely or partially at its reproduction cost, the estimated rate of return will be substantially lower than the rate based on capital actually invested. The substantial weight given to reproduction costs encourages management to use older equipment as long as possible, especially when there is an inflationary trend or when the evaluators are generous. In addition, the valuation and rate-of-return concepts generally lead to the "gold-plating" of facilities whenever the rate of return allowed is equal to or greater than the rate of return that the pipeline owner could realize by investing the capital elsewhere. This effect is particularly strong where the final demand is relatively inelastic and where the transportation facilities are owned largely by the shippers and/or receivers of the commodities. Both of these situations hold in the case of commodities moved by pipeline. Indeed, relating to a rate base both the rate of return on invested capital and the dividends that can be paid to the owners provides a dual incentive to have the rate base as high as possible, particularly since the principal owners of competing pipeline facilities share a common interest. The relatively high value (because valuation is based on reproduction cost) placed on old—that is, written-off—but still operable pipelines encourages owners to stretch the lifetime of these older facilities as far as possible in order to keep the true rate of return as high as possible. This is particularly attractive to pipeline companies that are independent of oil companies, such as Buckeye Pipeline, formerly independent and now a subsidiary of a railroad.

Some significant technological developments in pipelining have clearly been brought about by a combination of the character of regulation and the quality of management of other modes of transportation. Permitted by regulation to make rates collusively, U.S. railroads maintained rates on petroleum products through the 1920s and 1930s at a substantially higher level than was warranted by transportation costs of these products. Moreover, the ICC, when called upon to look at these rates, refused to judge them unreasonable. The excessively high rates profoundly influenced the structure of the petroleum production, refining, and distri-

27. *Petroleum Pipelines and Public Policy, 1906–1959* (Harvard University Press, 1967).

bution industries by safeguarding from undue competitive pressure less efficient integrated oil companies (that is, ones which controlled both supply and distribution, including pipelines). These high rates led directly to the introduction of pipelines capable of carrying refined petroleum products.[28] Certainly pipelines to carry such products would ultimately have been developed, but the innovation received a strong, early impetus from the high rail freight rates. No matter how the situation is viewed, transport sector resources were misallocated, at least for a while, because of the distortion of innovation in favor of petroleum products pipelines during the 1920s and 1930s.

The story is basically the same, though with some variations, for the transportation of solids by pipeline. The unwillingness of certain railroads to break the unified front of the industry and to modify their coal rate structure led directly to the establishment, between Cadiz and Cleveland, Ohio, of the first commercial solid pipeline in the United States. This pipeline was ultimately put out of business by railroads that offered a drastic rate cut. The rate reduction the railroads had to make to put the pipeline out of business was far greater than that which would have kept it from being built in the first place. Moreover, having opened Pandora's box, the railroads now face a proliferation of solids pipelines as the technology associated with this innovation is further developed and made economically and physically practical. Another example is the establishment of the LPG pipelines several years ago. The railroads failed to act even though an effective technological and pricing innovation was available to them in the form of lower rates if shipments were made in cars of 30,000-gallon capacity or larger.[29] It would seem a valid hypothesis that the railroads have failed to innovate partly because of the monopoly mentality induced by the regulation to which they have been subjected over many decades.

Regulation and Suppliers

The economic regulation applied to carriers has tended at best to dampen and at worst to distort the innovative propensities of suppliers.

28. Ibid., pp. 251–67.
29. Many, including the author, believe that a large part of the freight traffic the railroads have lost to other modes has shifted because the railroads abdicated their position without a fight, in the form of either lower rates, improved service, or both. Finished automobiles were in this category from 1930 to 1960 since the automobile rack car and associated tariffs could have been introduced decades

Technological innovation in these supporting industries is stifled by regulatory delay and by other aspects of regulation that reduce the profitability of innovation for carriers that are willing to take the risk associated with investing the necessary capital. An innovating supplier must not only persuade its carrier customers to try something new, but also rely on their ability to secure regulatory sanction to exploit the innovation if a price change is necessary to support its adoption. Obviously, the potential benefits of technological innovation emanating from the suppliers will be lower under regulation than would otherwise be the case.

Intermodal Competition

The effect—and the intention—of regulation in transportation has been to reduce competition among carriers, although even fewer of them might be operating if the ICC had not had the power to veto both intramodal and intermodal mergers. The effect on innovative propensities and possibilities of having more or fewer railroads within a narrow range is relatively unimportant. The number of independent carriers is already quite large and their interdependence requires industry-wide agreement for innovation to take place on a broad scale or indeed to be possible at all. The most significant competition in the transport sector as now constituted in the United States comes from intermodal, not intramodal, competition. Especially in the rail and highway modes, price (or rate) identity is a hallmark of the system. Anything that artificially handicaps one mode in relation to another violates an important overall objective of regulation—the optimization of resource allocation within the transport sector.

The agricultural commodities exemption provides an example of the distortion that results from economic regulation differentially applied in transportation. Carriers in other modes with which the railroads compete are exempt from regulation when hauling unprocessed agricultural commodities. If similarly exempt, rail carriers would devote substantial re-

---

earlier. This raises the question whether even enlightened regulation would improve the performance of railroads. The answer seems to be that it cannot do any harm and is probably a prerequisite to improved managerial performance in transport companies.

sources to developing rolling stock tailored to the mass handling of certain farm products and would devise innovative rate schemes and service patterns. They would then provide a valuable service and secure profitable traffic. The market and financial success of the Southern Railway's Big John cars specifically supports the conclusion that hauling exempt agricultural commodities is profitable when the transport system is properly designed for the purpose. The Southern's success, while certainly less dramatically profitable than it could have been (because of unwise regulatory behavior), was possible only because of the combination of technological, marketing, and pricing innovations.

Mergers and Public Policy

Public policy on mergers in the transport sector has become a matter of increasing concern. Attention has focused primarily on horizontal combinations, which in the case of rail and highway transport fall under the jurisdiction of the ICC. Commission approval is required for regulated highway carriers to amalgamate, for a railroad to gain control of, or join with, another rail carrier, and for railroads to acquire certificated carriers in other modes.

The impact of ICC constraints on technological innovation is difficult to assess in the case of same-mode mergers. Such mergers probably open up marketing or rate innovation possibilities that are unlikely if mergers are prevented.

Cross-modal amalgamations appear to offer a greater opportunity for introducing new technology. Intermodal movements in the United States would be much greater if unrestricted horizontal integration were permitted; that is, if joint ownership of carriers in several modes of transportation by a single firm were possible. The structure of the Canadian transportation sector bears this out. The Canadian National Railways and the Canadian Pacific Railway, the principal Canadian rail carriers, also operate extensively in other modes of transportation, either directly or through wholly owned subsidiaries. And far more innovation—technological and otherwise—in intermodal shipping is being attempted in Canada than in the United States, despite the smaller transport market in Canada. The effect of the size and character of a carrier on its willingness to undertake intermodal innovation has never been suitably explored; absolute size, intermodality, and length of haul may each play a role. Innovation in Canada has undoubtedly been spurred on by the opportunity offered to enhance profits by moving freight by the most efficient means, regardless of mode.[30]

Recent conglomerations and intramodal mergers by railroads have important implications for horizontal, intermodal mergers. If the proposed intermodal, integrated transportation companies are to provide satisfactory public service and substantially better allocation of resources in the transport sector, there must be competition among these integrated firms. Any merger in transportation that results in too high a degree of monopoly in one mode over high-density routes or in a geographical region makes it less likely that effective competition can arise among horizontally integrated transport companies. End-to-end railroad mergers should not per se undermine the attractiveness of the integrated transport company.

Conglomeration in the railroad industry drains resources and diverts managerial attention from the transport sector. This reduces the probability, first, that horizontal integration will take place across modal lines and, second, that resources—financial or managerial—will be available to carry out integration in a timely and efficient manner. The best interests of railroad stockholders in the short run may be served when managements seek the higher returns apparently available outside transportation. But perhaps other considerations are overriding. For example, returns generated by a well-managed, horizontally integrated transport concern might be comparable to those from investments outside the transport sector. In any event, carriers diverting resources from transportation may be acting against the public interest as embodied in the spirit, if not the letter, of the law.

Vertical Integration

In the United States regulatory agencies have paid little attention to vertical integration between carriers and firms that produce and market transportation hardware, despite the requirement in Section 10 of the Antitrust Act of 1914 (the Clayton Act) that the ICC monitor such relations.[31] In highway transportation, neither regulatory agencies nor the Justice Department has been concerned with vertical integration. One of the principal U.S. highway carriers, Consolidated Freightways, manufac-

---

30. Particularly dramatic is the relatively greater exploitation of unitization principles (including containerization) in Canada.
31. 38 Stat. 734; 15 USCA 20.

tures both tractors and refrigeration equipment for its own fleet and for other carriers as well. This arrangement seems to have benefited not only the parent carrier but also other truckers, including competitors. The evidence lies in the failure of other truckers to complain about the integrated firm and in their purchase of the equipment in spite of the availability of a broad range of alternatives.

Relatively little vertical integration has occurred in the railroad industry, although the current trend to diversification in American railroading raises the possibility.[32] Railroads already produce steel, chemicals, trailers, containers, and brake equipment.

A policy that apparently condones essentially unrestricted vertical integration through mergers can seriously affect the innovative propensities and performance of railroads. These combinations reduce competition for the business of the railroad (notwithstanding Section 10 of the Clayton Act), but the anticompetitive, anti-innovation impact might well be even broader through secondary effects. For example, a rail carrier might exert considerable, though subtle and lawful, influence on its principal traffic-interchange partners to favor its own railway supply affiliate, especially where materials are purchased under rigid specification, as is the case for many components of railway rolling stock that moves freely in interchange between the railroads.

The interdependence of the numerous railroad companies is one of the main reasons why vertical integration in the railroad field has such a deleterious effect on innovation. This contrasts sharply with the situation in the trucking industry, where, with few exceptions, equipment decisions are made independently by the carriers. For instance, tractors, trailers, and trucks with widely different equipment and operating characteristics can all be operated over almost any highway. This is not the case in railroading. A significant portion of the capital invested in railroads flows into rolling stock that must be standardized for interchange. But investment must also be innovative if it is to respond to shipper requirements and thus reverse the deterioration of the railroads' position in the marketplace.[33]

32. Several common carrier railroads are owned by large producers of railway materials, such as steel companies. In each case the railroad constitutes a small portion of the firm's activities and was acquired or established after the firm had been in its basic business for some time.

33. Equipment must be innovative whether railroads choose a general-purpose or a special-purpose approach to car supply. For example, large, lightweight open hoppers capable of hauling a wide range of commodities would be highly innovative in certain services without the carriers choosing the special-purpose approach.

For different reasons, interrailroad agreement and cooperation on the technology embodied in much railway investment other than rolling stock is also necessary. Economies of scale in the manufacture of most of the components of railway infrastructure require that a large number of rail carriers accept common versions of this equipment to induce manufacturers to develop and produce it at reasonable costs and prices. Indeed, the American Association of Railroads (AAR) establishes the boundary conditions and performance criteria for much of this equipment. Vertical integration is therefore a much greater threat to the well-being of the railroading industry than in other modes of transportation with less intramodal physical interdependence among its carrier constituents.

Considerable self-regulation is imposed through the AAR and other government-sanctioned intrarailroad committees and associations. While self-regulation probably has less impact on technological innovation than government regulation, it is nonetheless significant.

Much of the innovative drive in the transport sector comes from suppliers to transportation companies. The question thus arises whether the structure and market conditions in the transportation industry have had any influence on the structure and conditions of competition in the supplier trades. Unfortunately, very little data exist on which to base a firm judgment.

Highly concentrated conditions of supply are found in several key areas—for instance, locomotives, rails, signaling devices, and brake equipment. The concentration ratios provided in a 1966 publication of the Senate Subcommittee on Antitrust and Monopoly (summarized in Table 6-3) show an increase in concentration between 1954 and 1963 in most cases in which the data permit comparisons over time.[34] In every instance except one (nonpropelled ships) in Table 6-3, the dollar value of shipments increased throughout the period, in most instances markedly.[35] Unlike the transport equipment field, other producer goods industries experienced a decline in concentration, as is shown in the Senate subcommittee study.

All in all, the drift to greater concentration in transport equipment manufacture casts even more doubt on the wisdom of regulatory policy toward vertical integration. Carriers subject to regulation are integrating

34. *Concentration Ratios in Manufacturing Industry, 1963,* prepared by the U.S. Bureau of the Census for the Subcommittee on Antitrust and Monopoly of the Senate Committee on the Judiciary, 89 Cong. 2 sess. (1966), Pt. 1.
35. Ibid., pp. 231, 233, 234.

Table 6-3. Concentration Ratios in Selected Transportation Equipment Industries, Four Largest Firms, 1954, 1958, and 1963[a]

In percent

S.I.C.C.[b]	Class of product	1954	1958	1963
	Inputs to railroads			
3741–37422	Locomotives and parts	89	92	92
	Freight train cars, new	—	54	77
	Inputs to trucking			
37150–37172	Truck trailers	56	52	60
	Truck tractors, truck chassis, and trucks	77	78	81
	Inputs to water carriers			
3731–37311	Shipbuilding and repairing	43	48	48
	Nonpropelled ships (barges, and so forth), new construction	—	36	48

Source: *Concentration Ratios in Manufacturing Industry, 1963*, prepared by the U.S. Bureau of the Census for the Subcommittee on Antitrust and Monopoly of the Senate Committee on the Judiciary, 89 Cong. 2 sess. (1966), Pt. 1, pp. 231, 233, 234.

a. Percent of value of total shipments of each class of products accounted for by the four largest companies.

b. Standard industrial classification code.

vertically by acquiring firms in industries already exhibiting high and increasing concentration.

Neither the ICC nor the Justice Department has formally raised the question of the impact of vertical combinations on competition. The silence of the ICC is far more difficult to understand than that of the Justice Department, given the special relation between the commission and the railroad industry. Whether or not the framers of the Clayton Act saw an inherent danger in such combinations, Section 10 does give the ICC the power to review such arrangements ex ante and to police them ex post. Surely the Department of Justice and the ICC should consider such issues, particularly since once an undesirable vertical integration has taken place, it is difficult, if not impossible, to reverse.

Size of Carrier and Innovation

Up to now nothing has been said about how the effect of transport regulation on innovative capabilities and propensities differs according to the size of the carrier. In both rail and highway transport, regulation has worked to the disadvantage of the innovative small carrier. If forced to choose, the regulators are too often found on the side of the conservative, entrenched, often mistaken, larger producing units in the industry.

A classic example can be found in the history of the so-called Gartland rates advanced by the Chicago and Eastern Illinois Railway (C& EI) in 1957.[36] Briefly, the C&EI, alone among railroads, proposed to offer substantially reduced rates on water-rail-water and rail-water movements of coal, primarily as a counter to all-water movement via the inland waterways that parallel the C&EI's routes from southern Illinois and Indiana to Lake Michigan.[37] The large coal-carrying railroads, competitive either directly or indirectly with the C&EI, bitterly opposed the lower rates. Prominent among the C&EI's adversaries was the Chesapeake and Ohio, a large railroad by any measure and one of the largest coal haulers.

History since 1957 bears out the wisdom of the C&EI's move, whether made for the right reasons and with the correct vision or not. In the decade following ICC suspension of the C&EI Gartland rates, the railroads have seen a rise in substitutes for coal, primarily in the form of atomic energy, and in substitutes for moving coal by rail, principally in the form of the solids (slurry) pipeline and long-distance high-voltage electric transmission, by which coal is moved "by wire." If the railroads had published rates on a broad scale that were closer to the level of the C&EI Gartland rates at the time they were proposed, it is certain that substantially more coal traffic would have remained with the railroads. Furthermore, coal's emerging chief competitor, atomic power, would probably not have made the headway that it has.

The C&EI, which presumably wanted to remain competitive in both the transportation and the energy senses, was prevented from doing so by the ICC, acting largely at the behest of larger railroads. At best, these allies against change had the wrong vision of the coal market and therefore of the market for coal transportation. At worst, their motives reflect a strong bias against innovation in railroading.

Cases directly paralleling the C&EI Gartland rate situation can be found throughout the history of the regulated transport industries. One

36. *Chicago, Burlington and Quincy Railroad Co. et al.* v. *Chicago and Eastern Illinois Railway Co. et al.*, 310 ICC 349 (1960) and 315 ICC 37 (1961).

37. Whether the C&EI rate cuts were deeper than necessary to achieve the desired traffic results is of no concern here. In fact, the question is of doubtful relevance to the ICC. Was the commission seriously concerned that a cutthroat rate action by small, independent C&EI would undermine Norfolk and Western, Chesapeake and Ohio, Pennsylvania, New York Central, or the entire system?

can scarcely wonder at the persistence of the view that especially the larger, well-entrenched carriers—which often lack a planning conception —are, in their relationship with the regulatory agency, very much like prisoners who have fallen in love with their chains. Both regulators and regulated tend to be conservative, striving to maintain the status quo long after its inappropriateness to changed market and technological conditions has been demonstrated.[38]

If, as Adelman says, "a more competitive market conduces to the better use of existing technology and a better rate of innovation," regulation that is anticompetitive in its net impact must have a detrimental effect on the propensity to innovate and on the direction of innovation when it is undertaken. As Adelman points out, "strong competition makes innovation a necessary condition for profit,"[39] and where price identity among competitors is enforced, as it frequently is in transport, the attraction of demand-stimulating innovation is certainly reduced. On the other hand, even under present transport regulation, substantial resources will be devoted to cost-reducing innovation, but only if the resulting extra profits can be retained and, at least ultimately, shared with the owners. Certainly, in the transport industries that are regulated largely on a rate-of-return basis, as are the pipelines, the incentive to develop and invest in cost-reducing innovation is materially reduced if demand is relatively inelastic.

The existence of the rate bureaus (or more properly the rate-fixing bureaus) in the rate-regulated rail and highway fields[40] leads to coordi-

38. See M. A. Adelman, "American Coal in Western Europe," *Journal of Industrial Economics,* Vol. 14 (July 1966). Adelman shows the role American coal could have played in meeting Europe's energy requirements if transport costs had been lowered. The feasibility of lower rates is dramatically demonstrated by the Canadian Pacific's capture of a great quantity of new coal traffic in connection with a long-term purchase of Canadian coal by the Japanese ("Japan Sparks Western Coal Boom," *Canadian Business* [August 1969], pp. 26–33; and "Canada's Superport for Superships," *Business Week* [Sept. 27, 1969], pp. 120–22). The CP charge for shipping Canadian coal to the Pacific Coast is lower than the rate for shipments from Appalachia to the Tidewater area destined for Europe. Moreover, the Canadian venture requires substantial new capital facilities, including track, and is predicated on acquisition of new cars by the railroad, neither of which would be required for American shipments to Europe.

39. M. A. Adelman, "Technological and Scientific Problems in the Relations Between Europe and the United States," paper delivered at Turin, Italy, November 1967 (mimeographed), pp. 23, 8.

40. 62 Stat. 472 (Interstate Commerce Act).

nation in pricing and in output.[41] This means that the firms collectively act as monopolists. Rate bureaus were created under the sanction of the Interstate Commerce Act and other complementary legislation, another example of an anticompetitive regulatory rule that, on the basis of available evidence, was not examined by the framers of the legislation for possible anti-innovative effects.

The ICC and Innovation

Notwithstanding the failure of the authors of the Interstate Commerce Act to understand the high correlation between anticompetitive and anti-innovative forces, the evidence seems to indicate that the ICC has done the poorest job of all the regulatory agencies. Its policies, decisions, and indecision impinge on innovative drives and performance in the transport fields in which it has explicit jurisdiction or implicit influence. The commission has, for several reasons, performed especially badly where the railroads are concerned.

First, the Interstate Commerce Commission, alone among the federal transport regulatory agencies, has a mandate that is intermodal. The ICC is required to monitor, if not to regulate, growth and development in several modes of transportation and must adjudicate issues that cut across two or more modes. In effect, ICC regulation is intended as a substitute for both intermodal and intramodal competition. With such a mandate, and given the almost wholly legal orientation of the commission, it is not surprising to find it acting to preserve the status quo. Obviously this bias results in a regulatory environment that is anti-innovative at worst and that tends to distort management innovative drives at best. Indeed, the notion that the commission should "preserve the inherent advantage" of the modes under its jurisdiction[42] practically requires it to rule against innovative activity.

Second, in making decisions affecting the intermodal distribution of services, the ICC is frequently confused by differences in infrastructure costs arising solely from implicit government subsidies. For instance, when a pending rate case requires that the ICC compare the costs of per-

41. George W. Hilton, *The Transportation Act of 1958* (Indiana University Press, 1969), pp. 9, 31.
42. 54 Stat. 899 (Interstate Commerce Act).

forming similar transport jobs by road and by rail, the commission is confronted by a situation in which the rail carriers provide their own infrastructure and the highway carriers rely on governmental provision of most of the fixed plant they need. The problem of comparing these two situations is difficult, even for experts in cost accounting and economic analysis. Since it is largely legally oriented, the ICC is generally unable to cope with such situations.

Third, the comparative maturity of the several modes under the ICC's jurisdiction also makes it hard to regulate them in a way that will promote desirable innovation. In many mature industries like the railroads, excess capacity in fixed plant is substantial. This accentuates the difficulties of comparing the economics of such modes with others having a different level of maturity, durability of fixed plant, and growth rate.

Fourth, the sheer size of the commission would seem to engender an overly conservative, anti-innovative approach to regulation. With eleven commissioners, the ICC has the largest membership of any federal regulatory agency. On a five-man board, such as the Civil Aeronautics Board or the Federal Communications Commission, only three members at most need "see the light"; with the ICC, seven commissioners must be convinced of the necessity to encourage—or even accommodate—change. This practically guarantees that technological innovation and development will be thwarted far more often than would otherwise be the case.

Conclusion

Unquestionably, transport regulation has had an effect on innovation by the carriers and by suppliers. On occasion it has also influenced the shipper's desire to innovate, especially where transportation costs are a significant part of total delivered product price.

The magnitude of the effect of economic regulation on the incidence and character of innovation in the surface transportation industries is hard to assess with any precision. For one thing, regrettably few systematic attempts have been made by the regulatory agencies or other governmental bodies with a legitimate concern to relate past regulatory practices, policies, and decisions to managerial propensities to innovate in the regulated transport field and in relevant peripheral areas. Moreover, even if relevant data had been developed to shed light on the more direct relations, the indirect impact of regulation is often more important. This is especially true when heavy reliance is placed on the unregulated suppliers to provide the impetus for innovation to the carriers.

By and large, regulation in the transport sector is anticompetitive. Indeed, it was so designed. If, as seems likely, competition effectively breeds innovation of value to the entrepreneur and to the economy at large, diminution of competition is anti-innovative or, at best, distorts the innovative drives of management. In either case, the resources used in producing transportation are not allocated optimally. Minimum regulation and minimum interference with competitive forces would seem to promote the best innovative performance in the transport sector, especially since the technological possibilities for effective competition for most freight traffic have expanded so much that little noncompetitive traffic is left in the transport sector. And the traffic that remains noncompetitive is also threatened by technology, not only in the transport sector but also in other areas where substitutes for transportation, such as the development of ultrahigh voltage transmission systems to "move" coal by wire, are evolving.

A complete overhaul of the regulatory system for transportation by rail, highway, and to a lesser extent waterway and pipeline is unlikely in the near future. The best hope is that regulators will recognize that regulatory policies and specific decisions often—perhaps usually—have an impact on management's propensity to innovate. Regulators should be explicitly required to consider the relation between regulation and innovation, both in general and in each particular relevant case. To begin with, serious consideration should be given to implementing Recommendation 13 of the report of the Panel on Invention and Innovation, established by the secretary of commerce in 1965.[43]

A group should be established within the Federal Government to aid and advise the regulatory and antitrust agencies by performing such activities as:
(1) Developing criteria for helping these agencies judge the impact of antitrust and regulatory policies on invention and innovation.
(2) Systematically analyzing the consequences of past antitrust and regulatory activities in light of these criteria.
(3) Advising the responsible agencies on the probable consequences of proposed policy changes affecting invention and innovation.
(4) Providing technological forecasts as an additional factor for antitrust and regulatory planners to weigh in their policy formulations.

Recommendation 14 makes it clear that the regulatory field has too

43. Panel on Invention and Innovation, *Technological Innovation, Its Environment and Management* (Government Printing Office, 1967).

long been a stronghold of those with legal training, almost to the exclusion of others:

> To enable the antitrust and regulatory agencies to give greater attention to questions concerning technological innovation, their staffs should be strengthened by increasing the number of personnel who have a deep understanding of economic and technological development.

The transport regulatory agencies have, without exception, operated with little or no competent economic and engineering counsel. No significant improvement in the sector's innovative performance is likely to occur as long as economic regulation remains almost exclusively the province of lawyers.

Innovative behavior is forward-looking, whereas legal thinking tends to be conservative. To superimpose the continuous, orderly, and evolutionary legal perspective upon what must be a discontinuous and revolutionary process of technological innovation is to limit technological development in the regulated industries. This is dramatically illustrated by the necessity to substitute a dynamic concept of "inherent advantage" for the static one that dominates regulation in surface transportation. Regulatory agencies tend to treat as permanent an advantage one mode may temporarily have over another, although technological change may quickly overcome this.

In the long run, the innovative performance of the transport sector can best be improved by gradually eliminating economic regulation and probably by permitting, if not encouraging, horizontal integration. Competitive integrated transport firms would offer the public a range of precisely specified transportation services, which would be produced by using the mode (or modes) that accomplish the task with the least expenditure of resources. Greater skill in matching modes and methods with the shippers' needs would generate larger profits for the horizontally integrated transport concern. Each firm would be encouraged to innovate both technologically and nontechnologically to the extent necessary to maximize its own profits. The firm would tend to behave more as though it were not regulated. In an environment characterized by vigorous competition between strong, integrated transport firms, the workings of the antitrust laws should assure enough competition to avoid the dangers and inequities that would otherwise arise, given the cost structure and the cost-price relationships in the industry.

CHAPTER SEVEN

Summary and Conclusion

William M. Capron
and Roger G. Noll

PRELIMINARY VERSIONS of the preceding chapters were the basis of the conference held at the Brookings Institution. This chapter discusses the main points of each chapter, the principal issues raised during the conference, the conclusions reached by the conferees on those policy issues about which a clear consensus emerged, and some issues left unresolved at the conference. Since the conferees were not polled and since no effort was made to develop a consensus, the following discussion reflects only the authors' judgment of the views of the conference participants. One objective of this book being to spotlight topics needing more research, the authors also assess in this chapter the present state of professional understanding of the process of innovation in regulated industries.

Conceptual Framework

The purpose of the paper by Fred Westfield was to provide an analytical foundation for examining the effect of regulation on technological change. To make the problem conceptually and mathematically tractable, Westfield narrowed the scope of his inquiry to the impact of three regulatory practices—controlling the rate of return on capital, restricting the profit markup over costs, and specifying ceiling prices—on three specific types of innovations—Hicks-neutral, Harrod-neutral, and the capital-augmenting obverse of Harrod-neutral.

Westfield assumes that the primary objective of the regulatory agency is to keep price lower and output higher than would occur if the industry were monopolized by a single, profit-maximizing, unregulated firm. To use Westfield's terms, regulation seeks to establish a "profit-permissibility" schedule that leads a monopolist to decide to produce more output and to sell at a lower price than if he could pick the optimal point on his unconstrained "profit-possibilities" schedule.

Westfield's results show that all three regulatory methods for controlling price and output provide some incentive to take advantage of each type of technological change as long as the demand for the product of the regulated firm is elastic (that is, as long as a price cut increases the total dollar volume of sales). Westfield confirms and extends the results of Averch and Johnson[1] by showing that the bias toward increasing the capital–output ratio in firms subject to rate-of-return-on-capital regulations extends to a preference for capital-using innovations. Further generalizations about Westfield's results are difficult to make, for the magnitude of the effect of technological change on profits, and thereby on the incentive to innovate, depends on the elasticities of the demand curve and the production function. No particular type of price-profit regulation produces uniformly greater or lesser incentives to innovate for all kinds of demand curves, production functions, and technological change.

Westfield recognizes that the model he has developed is not complex enough to deal with many important empirical and policy issues. To rigorously develop unambiguous results, he used a comparative statics model built on highly simplified assumptions.

In discussing the Westfield paper, the conferees dwelt at length on the sensitivity of his results to his assumptions and on the alternative, more complex assumptions they believed were required if important issues beyond the scope of the paper were to be explored.

One of the issues discussed was the conventional hypothesis that regulated firms behave as if they were attempting to maximize profits. An alternative hypothesis with considerable support among the conferees is that the protected position of regulated firms creates passive, "satisficing" behavior. This view sees the decisions of regulated firms as largely reflecting historical precedent and bureaucratic inertia as long as profits, sales, and technical sophistication increase at least as fast as some normal, acceptable benchmark rate. Only substantial pressure arising from

1. Harvey Averch and Leland L. Johnson, "Behavior of the Firm under Regulatory Constraint," *American Economic Review,* Vol. 52 (December 1962), pp. 1052–69. See also Chapter 1 (pp. 4–5) in this volume for a brief description of the Averch-Johnson effect.

some cataclysmic event, usually external to the firm, causes a firm with this behavior pattern to change fundamentally its long-term performance path. An important consequence of the satisficing hypothesis is that the existence of regulation is far more important than the choice of regulatory mechanisms employed; a firm with acceptable, or satisficing, profits may well be concerned primarily with the global efficiency implications of a capital investment or a new technology rather than with the profit implications, and thereby not follow the types of biases predicted by the Averch-Johnson effect or in the Westfield paper in choosing among alternative technologies.

Another approach to understanding the effect of the regulatory process suggested during the conference is to focus on the interface between regulators and regulated, perhaps using a bargaining or bilateral monopoly model, rather than on the relation between producer and customer, as in the Westfield or satisficing models. This approach requires plowing much as yet untilled theoretical and empirical ground, including (1) specification of the objectives of regulators and regulated vis-à-vis each other, (2) development of a theoretical model of the determinants of bargaining strength for both sides, and (3) establishment of the permissible range of bargaining solutions and of how that range is affected by the characteristics of the regulatory process.

The conferees also discussed the view that regulation can be adequately characterized as the imposition of "public interest" constraints by a regulatory authority on the behavior of a private firm, an assumption implicit in profit-maximizing, satisficing, and bargaining models. The extreme opposite view, held by some conferees, is that regulatory agencies "crawl in bed with industry," often acting as no more than conduits for the interests of regulated firms and giving a legal sanction to cartel-like behavior that, in the absence of regulation, would never be permitted by either the political or the judicial system. Such regulation is very expensive—not only does it fail to make serious inroads against the monopoly profits of public utilities, but it generates the additional direct costs of regulation for both industry and government. Furthermore, to the extent that regulation is ineffective in limiting profits, the case for the profit-maximizing hypothesis about firm behavior is more plausible, and effects predicted by Westfield and Averch-Johnson are more likely.

No clear consensus emerged on the alternative hypotheses about either firm or regulatory agency behavior. Perhaps the most common view of firm behavior was that the profit-maximization hypothesis, although an oversimplification, is useful and reasonably valid, lending credence and explanatory power to analysis along the lines presented in the Westfield paper. While the conferees differed in their assessment of the behavior of regulatory agencies, they were nearly unanimous in the belief that the objectives of regulators are more complex than simply to constrain monopoly profits by reducing prices and increasing output in a static equilibrium sense. For example, agencies probably regard higher profits through innovation more favorably than higher profits through price increases, which, ceteris paribus, provides an incentive to do more than the optimal amount of innovating and to camouflage price increases as service innovations. While the conference produced no definitive results concerning the objectives of regulatory bodies or the ability of regulators to achieve their own ends, or of the importance of the effect of these considerations on the performance of the regulated firm, the discussion did point to the fact that too little research has been directed at this aspect of the regulatory process.

Another complexity examined only briefly by Westfield but of profound importance in the eyes of the discussants was the effect of the "regulatory lag" on the incentive to innovate. If regulatory agencies quickly perceive the effects of innovation on costs and profits and rapidly adjust profit and price regulations accordingly, regulated monopolies have much less incentive to innovate. Several participants believed that only a dynamic model, considerably more difficult to construct and to manipulate than Westfield's, can correctly show the effect of regulatory lag.

One participant suggested that the Westfield paper considered too limited a range of technological change, being concerned only with innovations that reduce factor requirements for producing a unit of output of a given previously produced commodity. Other types of technological innovations include introducing a new commodity or service that is only slightly, if at all, different from previous products of the firm; devising ways to substitute one factor of production for another in proportions more favorable than permitted by the old production function; and improving product quality with no change, or perhaps even an increase, in unit production costs. The model might also be able to take account of profitable technological regressions; that is, changes that raise costs but, because of regulatory rules, also raise profits.

Another conferee believed that the model should explicitly include research and development (R&D) costs and other costs of innovation. Many important questions relating to innovation costs can be raised.

How is the amount, content, and emphasis of the R&D conducted by regulated firms and their suppliers affected by regulation in general and by
the particular regulatory mechanisms employed? Does regulation, particularly if the rate of return is controlled, encourage R&D oriented toward
market expansion or the development of new markets, as the Averch-
Johnson analysis implies? Can government effectively offset R&D and innovation biases due to regulation by selectively supporting certain types
of R&D? How has past government-supported R&D affected technological change in the regulated sectors?

Several conferees questioned Westfield's separation of the three regulatory controls—ceiling prices, markup limits, and rate-of-return targets
—into discrete, alternative policy vehicles. In practice, most regulatory
agencies use a combination of these devices, as well as several others.
While rate-of-return control is commonly the principal device used in
regulating prices and profits of electric utilities, transportation, and communications firms, the federal agencies responsible for regulating these
sectors also regulate some prices directly. Most regulated firms do not
supply a single product to a single class of customers; instead they sell
several related products in a number of different markets, separable by
location and type of customers. Regulatory agencies, especially the Interstate Commerce Commission (ICC), tend to be as concerned about
price structure as about price level or rate of return. In fact, the rate of
return is often no more than the benchmark by which price level and
structure are judged. The extent to which regulated firms are allowed to
discriminate in price among users can significantly affect the firm's reaction to the possibility of undertaking various innovations.

The conference participants agreed that Westfield's analysis was careful, complete, and rigorous. As Westfield noted, much of the discussion
was about the paper he had not written, rather than the one he had. The
conferees agreed that a departure from the simplicity of Westfield's
model would have meant sacrificing clear, rigorous results. Indeed, a
more general approach might have been able in the end to say very little.

Electric Power

William Hughes explores in detail the developments over the past two
decades in the technology of generating and transmitting electricity. The
minimum size of the power system large enough to take full advantage of

scale economies in generating equipment has increased steadily and significantly. The bulk-power industry at present includes approximately
100 major systems with separate managements. Most of these systems
are too small to take advantage of available scale economies in power
generation. Hughes estimates that potential scale economies unrealized
because of the present structure of the industry account for 4 to 10 percent of wholesale power costs. Furthermore, because the market for the
most efficient generating equipment is smaller than would be optimal,
progress in bulk-power technology has probably been too slow. The
magnitude of these efficiency losses is not trifling; even if the possible
effect on technological change is not taken into account, unrealized scale
economies alone account for several hundred million dollars annually.

Hughes's research into industry costs and technology leads him to con-
clude that the bulk-power industry should contain no more than twenty
to thirty major planning units (each either a single company or a tightly
integrated power pool of independent companies). More than this degree
of concentration, argues Hughes, should be discouraged on a variety of
grounds (keeping equipment markets competitive, providing performance standards for intersystem comparison, limiting the size of individual companies to the minimum dictated by efficiency to protect regulators from being faced with regulating a very small number of politically
and economically very powerful firms, maintaining maximum flexibility
should technological change prescribe further restructuring the industry,
and so forth). Hughes advocates encouragement by regulators of mergers until the twenty-to-thirty-system industry is attained. He acknowl-
edges that power pooling, joint ventures, and other coordination devices
offer alternatives to mergers for achieving a better industry structure;
however, he believes that experience with these alternatives shows that
decision by committee is usually less efficient in the power industry.
Ownership is now so fragmented and uneven, according to Hughes, that
a great deal of reliance must be placed upon mergers if the gap between
actual and potential performance is to be closed.

The evidence amassed by Hughes convincingly shows that technological change in the electric power industry has generally taken the form of
steadily uncovering greater scale economies in successively larger sizes of
generating units, or, as Hughes expresses it, of pushing out the scale
frontier. Except for the recent development of using nuclear reactors as a
source of heat, no single, dramatic innovation has substantially changed
generating efficiency or cut the costs of long-distance transmission of

electric energy. Instead, progress has occurred through a series of incremental improvements. In generating equipment, these changes have been manifested by steady increases in the scale frontier and in temperature and pressure. In transmission, examples of incremental gains include new extra-high-voltage long-distance cables, circuit breakers and transformers, and load-frequency control devices.

The nature of electric power generation makes the industry very capital intensive. Technological change has increased capital intensity somewhat by substituting capital for fuel. (Direct labor costs have long been insignificant, dominated by capital and fuel.) The marginal increases in the capital–output ratio caused by technological change have been more than offset by savings in fuel costs; in the absence of these technological improvements, electricity prices would be significantly higher.

Participants in the Brookings conference were particularly interested in the implications of Hughes's finding that technological change in electric power has been consistently in the direction of discovering ways to make larger and larger generating units. One reason mentioned at the conference is that development efforts by generator and turbine producers have reflected the rapid, steady, and predictable growth in demand for electrical energy. But some conferees pointed out that, according to Hughes, most power is generated by firms too small to take advantage of the technological advances being made, regardless of how fast demand is growing; why has technological change ignored the potential demand among these producers for more efficient generating systems of smaller than maximum scale? The suggestion was made that the character of technical advancement provides evidence supporting the Averch-Johnson effect in that power-generating firms show a marked preference for ever-larger units which, all other things being equal, require the power system to invest in more reserve capacity.

The question arose in the conference discussion whether the developments described by Hughes could properly be considered as technological change since most of the changes could be described as movements to an untried but known region of a single production function. The concept of technological change is fuzzy, involving technical, organizational, and behavioral dimensions. Conferees supporting the notion that advancements in the efficiency of power generation are properly identified as technological change cited three reasons: the need for developmental experimentation to achieve increased unit size, the uncertainties surrounding the cost and efficiency performance of larger units before they were put into use, and the reorganization of both generating plants and the relation of plants to the power system required to take full advantage of larger units. To regard the progress in the power industry as something other than technological change is, by implication, to require that technological change embody some dramatically new scientific knowledge. Most conferees believed such a definition of technological change to be overly restrictive.

The conferees generally agreed that the Federal Power Commission (FPC) has had little effect on the power industry. Since the ever-increasing scale of generating units has led to progressive reductions in costs and prices, FPC regulation has taken what one participant described as a "passive and benign posture." The FPC did assume a positive role in conducting the National Power Survey, completed in 1964, which encouraged the development of more extensive pooling arrangements and other institutional changes rationalizing the structure of the industry. Some specific steps toward coordination have subsequently been actively supported by the commission, especially since the famous 1965 power blackout in the Northeast.

Federal regulation of prices in the power industry is confined to wholesale transfers of power. Most price regulation is by state and local government agencies. The conferees agreed that the magnitude and the direction of the effects of price regulation on technological change in the power industry are unclear. Most agreed that regulation has kept some pressure on electrical utilities to lower prices since most regulatory agencies publicly push for passing along cost reductions, at least in part, to users. Yet even without regulation, utilities have incentives to keep electricity prices low. New capacity is available only in large, discrete "lumps," the size of which is growing because of technological change. A single new generating station can increase capacity considerably; low rates bring capacity utilization up to the level of lowest unit cost more quickly. Moreover, the demand for bulk power, an important segment of the power market, is elastic. Major industrial users and smaller local distribution systems that are primarily power retailers both have the option of generating power themselves. Many industrial consumers can shift to alternative energy sources if electricity prices do not remain competitive. In the long run, consumer demand may even be elastic, although it is clearly inelastic in the short run. At various price thresholds a major share of consumer demand will shift from other energy sources to electricity (for example, by changes in the energy base of home appliances and heating systems). In many areas competition is particularly intense between electricity and natural gas for major shares of both the consumer and industrial market. All of these factors provide a strong incentive for electric utilities to adopt new technology that permits price reductions.

Hughes identifies a few large utility systems that, together with the major equipment suppliers, have been the technological pioneers. Since the principal technological changes have been connected to increases in equipment size, a firm must expect a large absolute increase in power demand to justify exploiting the latest scale economy. Furthermore, without close coordination with other systems, an electric utility cannot safely lodge more than a small percentage of its peak market demand in a single generating unit. Reserve capacity requirements are larger for a system using a few large units than for an equal-capacity system using many small units. In addition, generating plants suffer serious efficiency losses if operated far below optimum capacity, creating problems in providing power efficiently with a few relatively large units at times other than peak periods. For all of these reasons only a few of the largest utilities are consistently in a position to exploit the latest technological advance along the scale frontier for generating equipment.

Hughes perceives a fairly stable follow-the-leader relationship in the power industry, suggesting that in regulated industry, as elsewhere, some firms have more enterprising management than others and that once a firm has become a technological leader it is apt to continue in that role. Since the technological leaders are generally the larger firms, separating the effects of management style and of company size is difficult. During the discussion the question was raised, and left unanswered, whether, among large power systems, technological leadership was correlated with the nature of local regulation.

Hughes forecasts that power generation technology will continue to favor ever-increasing unit size. Nuclear plants promise even greater scale economies than plants using fossil fuels. Light-water reactors, currently used in nuclear plants, have already demonstrated substantial economies of scale, and power plants built around fast breeder reactors, which will become operational in the 1980s, will have to be even larger to be more efficient than plants using present reactors.

The discussion of nuclear power draws attention to a peculiar characteristic of the electrical energy industry. Though technological change has occurred more rapidly in the electric power industry than in most other sectors of the economy, the industry has done almost no R&D. Instead, it has relied on the equipment industry and other suppliers and on the federal government to support R&D. Only in transportation can one find a regulated industry with a comparably small R&D effort.

Common Carrier Telecommunications

William G. Shepherd's chapter presents seven hypotheses about innovation by regulated monopolies such as communications common carriers in general and the dominant communications firm, the American Telephone and Telegraph Company (AT&T), in particular. He evaluates these hypotheses by drawing inferences from the historical performance of AT&T under different conditions of regulation and external competition.

One class of hypotheses proposed by Shepherd relates to the behavior of AT&T as virtually a monopolist in domestic communications services, irrespective of the effects of regulation. Shepherd argues that monopolists, in general, will tend (1) to be less cost-conscious than firms in more competitive industries, paying less attention to cost-reducing innovations and the efficiency with which R&D is carried out; (2) to underinvest in R&D except when potential competitors threaten to enter some part of the market, in which case monopolists may overinvest in certain types of R&D to preclude entry; (3) to favor technologies that protect the monopoly position of the firm by preserving or extending scale economies that create natural monopoly or by integrating the production of the monopolized products so thoroughly that potentially competitive parts cannot conveniently be separated and opened for entry of new firms; and (4) to enter unprofitable markets as a device to capture other, more profitable markets, or to foreclose entry by a new firm that might eventually compete in the profitable markets.

According to Shepherd, regulation in communications has reinforced some of the existing tendencies toward inefficiency in technological change due to monopoly and created incentives for additional inefficiency attributable to regulation alone. He argues that both regulatory actions and antitrust settlements have buttressed AT&T's ability to foreclose potential competition. The FCC's accession to AT&T's prohibition of "foreign attachments" on the switched network, eventually reversed in 1968, and the consent decree preventing competition between Western Electric and other equipment producers are both cited as examples of how federal government policies have strengthened the monopoly position of AT&T in communications.

Shepherd cites rate-of-return regulation by both the FCC and local regulatory agencies as a potentially important influence on the innovative performance of the industry. Rate-of-return regulation induces firms to favor capital-intensive innovations and to find entry into certain unprofitable markets even more attractive, both corollaries of the Averch-Johnson effect.

The historical evidence about the performance of the communications industry, according to Shepherd, suggests, but does not prove, that all but one of the above biases and inefficiencies are operative in communications. The exception is the possibility of a failure to be "X-efficient"— that is, to be inefficient in resource use—in R&D. While the variance in performance among Bell System operating companies suggests that some may produce communications services inefficiently, no evidence exists that Bell Telephone Laboratories and Western Electric Company, which perform nearly all of the R&D for the Bell System, are inefficient. Most of Shepherd's evidence in support of his remaining hypotheses is based on the corporate position of AT&T on three major issues in communications policy: the possibilities for competition and deregulation afforded by microwave developments in the late 1940s and again in the early 1960s, the prohibition against foreign attachments to the switched network, and the choice of technology and ownership structure for communications satellites.

The Bell System has twice faced competition in microwave transmission. In the first instance, competitors threatened to construct a microwave system to compete with Bell's coaxial cables; AT&T warded off this threat, according to evidence cited by Shepherd, by reversing an internal policy against microwave development and undertaking a crash R&D program to beat competitors to the market and preserve the company's monopoly position. One conferee mentioned the early reluctance of AT&T to develop microwave as consistent with both the lethargic innovative-propensity and capital-intensity hypotheses. The later reversal was possible only because of the enormous amount of basic research on microwave that had been achieved at Bell Laboratories—cited as evidence for the entry-foreclosing R&D hypothesis—and, according to Shepherd, was consistent with the hypothesis that AT&T will act to preserve its monopoly. A similar microwave issue arose in the 1960s when independent firms began offering private line microwave service in a few high-density markets. By its own admission, AT&T responded by offering service at less than fully remunerative prices, indicating the strength of AT&T's resolve to retain a communications monopoly.

The foreign attachments prohibition is used by Shepherd to illustrate several of his hypotheses. First, owning, rather than selling, terminal devices adds to AT&T's rate base, increasing allowable profits. Second, the foreign attachments prohibition forecloses competition in producing terminal devices. Third, by owning attachments a potential opening wedge for competition in providing communications services is prevented. Finally, the fact that AT&T developed but did not, until forced, introduce a device to prevent technically inferior foreign attachments from damaging the entire switched network is cited as additional evidence for all of these points and for the hypothesis that an imbalance exists between R&D and actual innovation which insures AT&T against potential competition.

In the case of satellites, AT&T pushed the development of random orbit satellites (with much higher capital costs than the geosynchronous type) and insisted that they have partial ownership of some component of the satellite system instead of being relegated to the role of leasing satellite circuits. Furthermore, in place of further expansion of the satellite system and despite much higher costs for cables, AT&T promoted the laying of another transatlantic cable in the late 1960s. Shepherd argues that all of these positions are consistent with AT&T's hypothesized propensity to extend its rate base by choosing capital-intensive technology.

Shepherd's primary policy recommendation is that communications regulators should seek more opportunities to introduce competition into the communications sector since, he argues, competition is more effective than regulation in promoting proper innovative performance. To this end he proposes that serious consideration be given to policies that would permit freer entry into microwave transmission, keep the domestic satellite system separate from AT&T, and either expunge or modify the consent decree to open at least some part of the communications equipment business to competition.

Shepherd also believes that the wisdom of the continuous surveillance technique of the FCC in regulating communications is open to question. While continuous surveillance may preserve the FCC's admittedly scarce resources, the absence of formal hearings prevents the establishment of an open public record on the performance of the industry. Consequently, outside observers, whether academic researchers or government officials, can draw only limited inferences about the effectiveness of regulation and the progressiveness of the dominant firm. Formal hearings also afford the regulatory agency some protection against an appeal by communications firms to the political system, where, in the absence of concrete evidence unfavorable to them, they have more influence than their regulators.

The conference participants discussed at length the extent to which Shepherd's evidence is open to alternative interpretations. Some questioned his interpretation of the facts. Did the performance of AT&T in microwave development after World War II reflect a radical switch in company policy, or simply the normal transition from slower, less costly basic research activities to more easily completed but expensive system development and implementation stages? Did AT&T exhibit inordinate myopia or corporate bias in promoting random orbit satellites, or did Hughes Aircraft accomplish a near-miracle, astonishing the entire aerospace industry, by developing geosynchronous satellites so quickly? Can AT&T be faulted if the Defense Department, the State Department, and the FCC all pushed for another expensive transatlantic cable because they believed the cable and the preservation of the cable-producing industry to be required for national defense and foreign policy purposes? And considering comparative risks and costs of failure, are cables really more expensive?

Others argued that even if the cases cited by Shepherd did show suboptimal innovative behavior by AT&T his hypotheses were neither necessary nor sufficient to explain AT&T behavior. On the one hand several of Shepherd's hypotheses can be dropped without reducing the extent to which the remaining hypotheses explain the evidence he cites; for example, the monopoly-preservation and capital-intensity hypotheses alone could account for all the behavior cited by Shepherd. On the other hand hypotheses he does not consider are also consistent with the evidence; for example, AT&T may simply act so as to maximize sales revenue, subject to the constraints that profits exceed some minimum acceptable amount and that the firm always seek to operate the existing plant (using existing technology) efficiently. A firm with these objectives could be expected to be unenthusiastic about pursuing innovations that reduce costs in markets with inelastic demand (which AT&T believes to be the case in business services and international communications) and about the prospect of losing parts of its market to potential competitors.

Considerable discussion focused on why the theoretical propositions about the characteristics of innovation in communications have not yet been tested conclusively. One reason, indicated by Shepherd and made more explicit in the conference discussion, is that theory suggests much more bias in the direction and composition of innovative activity than in its level. While Shepherd argued during the conference discussion that the rate of technological change in the communications sector may be less than indicated by commonly used methods of measuring productivity change, he and the conferees generally agreed that the communications sector has been rather progressive, making it difficult to separate what might have occurred from what has actually happened.

Several aspects of regulation by the FCC that are neglected in theoretical models might effectively work to counterbalance the effects of anticompetitive policies and rate-of-return control. One offsetting factor may be the policy of continuous, informal surveillance which Shepherd attacks as shielding the industry from public scrutiny. The counterargument is that AT&T, wishing to avoid the expense and publicity of threatened formal hearings, may be willing to bend company policy even further in the directions desired by the FCC than it would if forced into a formal hearing and possible subsequent litigation of the outcome. Another offsetting factor is the FCC's practice of occasionally investigating the economic justification of major investments by common carriers before permitting their inclusion in the carrier's rate base, in principle foreclosing the possibility of investments in outmoded technology, unremunerative markets (except where consistent with FCC policy), or innovations substituting capital for other factors at a net social cost. Shepherd pointed out that the effectiveness of the latter policy is clearly limited by the size and quality of the FCC staff, which cannot match the army of analysts employed by AT&T.

The conferees generally agreed that the degree of vertical integration and the size of AT&T probably are greater than necessary. One conferee noted that AT&T's case for vertical integration, based on the presumed need for experience in all phases of the communications business, does not logically lead to the conclusion that 85 percent of the common carrier market should be served by AT&T operating companies. Perhaps much could be gained by spinning off several AT&T operating companies to form another large, vertically integrated communications firm. The Bell System, a larger General Telephone and Electronics Corporation, and a third large communications firm created from independents and pieces of AT&T might introduce some competition in certain areas of communications and would certainly provide comparative performance information useful in establishing regulatory policies.

The Civil Air Transport Industry

Almarin Phillips traces the development and adoption of new aircraft by commercial carriers since 1932. Flying has been revolutionized in the past three decades, becoming faster, safer, and cheaper (relatively for *all* commercial passenger service, and absolutely for long-haul fares) with each major innovation in aircraft models. For example, the Douglas DC-3, introduced in 1936, had a seating capacity of 21, a cruising speed of 175 miles an hour, and a range of 850 miles; her later sister ship, the DC-8, introduced in 1959, seated 132 and had a speed of 550 miles an hour and a 4,000-mile range. The DC-3 had dramatically lower operating costs than planes in service at the time it was introduced. Planes in the early 1930s cost 7–10 cents per seat-mile to operate. Operating costs were about 3.25 cents a seat-mile for the DC-3 and less than 1.5 cents for the DC-8 (all expressed in 1954 dollars).

Aircraft manufacture is an unregulated industry composed of several rival firms, none of which are connected with the regulated air carriers and all of which are major defense contractors. (This separation has not always existed. Prior to the Air Mail Act of 1934, many airlines were divisions of aircraft manufacturing firms. Congress, in passing the statute, reflected concern over vertical integration in the industry and expressly forbade manufacturers to own airlines. This policy has been maintained ever since.)

What explains this record of steady cost reduction combined with dramatic quality improvement? The conferees agreed with Phillips that the federal government has played the dominant role in the development of basic aviation technology in this country. In the years during and since World War II in particular, Defense Department expenditures for military aviation R&D have been massive. Even before that, federal support was important, particularly from the National Advisory Committee for Aeronautics (NACA), the predecessor of the National Aeronautics and Space Administration (NASA). Many of the specific developments that have influenced the pace and direction of civilian aviation have unquestionably come out of the military programs.

Some disagreement did emerge at the conference among those knowledgeable in the field of aviation on a much more limited and specific question: How important to the development of civilian air transportation has been the conversion to use by air carriers of vehicles originally designed for the military? Conferees contending this has been significant pointed to the direct relation between early postwar civilian aircraft and planes developed during the war for military purposes. They also emphasized the close relation between the KC-135, a jet tanker developed for the Air Force, and the first civilian jet, the Boeing 707. Conferees holding the opposing view pointed out that Boeing began work on the 707 prototype in 1952, two years before the Air Force ordered the KC-135. The large government order enabled Boeing to offer the 707 at a lower price than it otherwise could have, but the plane is *not* an example of a commercial adaptation of a military development. One clear case of a plane first developed for military use and later adapted for commercial service was an earlier Boeing plane, the B-29 of World War II fame, but some conferees quickly pointed out that the civilian version—the Stratocruiser—was not a commercial success for either the manufacturers or the carriers.

The conferees who believed that many of the significant developments have come in direct response to the needs of civil aviation, and not as a result of military R&D, called attention to such improvements as cabin pressurization, noise suppression techniques, and the turbofan engine, and to the intense interest of civil air carriers in innovations that reduce costs. For example, longer time between overhauls and lower operating-when-empty costs have decreased maintenance and direct operating costs. Many specific model changes, particularly "stretching" an existing aircraft, are clearly the results of competitive pressures on the major airplane manufacturers and the carriers they supply.

The conference participants generally agreed that, with the exception of the jumbo jet development (for example, the C-5A), the major emphasis of military aviation today is probably much less applicable to future adaptation to civil aviation. (Other possible exceptions noted were the development for military uses of helicopters, STOLs—short takeoff and landing—and VTOLs—vertical takeoff and landing.) Also regarded as important were further innovations to improve safety and reliability in civil aviation and the recent emphasis on the environmental effects of aircraft, both of which are of much less interest in military applications and therefore call for efforts focused on civilian aviation.

Progress in military subsonic jets led rapidly to civilian use of similar aircraft because these planes were bigger, faster, and cheaper to operate than propeller-driven planes. The intense debate in this country during the past several years about the supersonic transport has arisen because it seems clear that (at least in the foreseeable future) supersonic planes will be *more* expensive to operate than the planes they will replace. Indeed, the difference in operating costs will be even greater than is suggested by present subsonic jet costs per seat-mile, since the so-called jumbo jets (subsonic) are expected ultimately to have much lower seat-mile costs than the jet aircraft now in commercial use.

The conferees agreed that the principal impact of regulation on technological change in the civilian air transport industry in this country has been an indirect one. Regulation has prohibited price competition among the air carriers and has thus channeled the rivalry among them toward service improvement. Some conferees suggested that innovation actually has been too rapid or of the wrong kind. One example is the pace at which jet transports were introduced into civil air transport in this country. In an oligopolistic industry in which price competition is suppressed, when one competing carrier adopts a faster and longer-range aircraft, rivals flying the same routes are forced to do the same as quickly as possible. This is most dramatically illustrated in transcontinental routes.

Several influences led to the introduction of the jet. In the first place, as was noted above, the feasibility and economy of civil jet aircraft were directly demonstrated by changes in military aircraft in the late 1940s and early 1950s. At the time when the initial decision to develop civil jet aircraft was made, the leading contender on the technological horizon was the turboprop aircraft. One participant noted that many experts believed the turboprop to be a much more efficient aircraft than the jet could possibly be and that the carriers should have purchased turboprop planes. Some believed that the choice actually made implied a decision on the part of both the aircraft producers and the air carriers to leapfrog what would have been an appropriate stage in the technological evolution of civil aviation. In retrospect, the move to jets may not have been premature, since, for one thing, the operating costs turned out to be considerably below the preliminary estimates.

The oligopolistic structure of the industry, combined with the suppression of price competition by the regulatory agency, does create some problems. Specifically, and again by using the introduction of the jet as an example, attention was drawn to the equipment boom that hit the industry in the 1950s, when carriers scrambled to acquire jet aircraft in order to remain competitive in both speed and range. Had price competition among the airlines been allowed, some firms might have kept slower propeller-driven aircraft in service, with fares below the jets'. Regulation ruled out this choice, except for a short time for the nonscheduled carriers. The cost of jets per plane was affected in opposite ways—the large number of orders permitted manufacturers to take advantage of scale economies, but the intense pressure to produce as many jet aircraft as possible in a very short time (which necessitated paying overtime rates for labor) created production inefficiencies. On balance, the well-documented "learning curve" effect on aircraft fabrication costs was probably stronger, and the large orders probably resulted in some reduction in the cost per plane.

Although Phillips regards the direct influence of regulation on technological change in aircraft as minor, he recognizes that regulation has indirectly influenced the direction and composition of development. Planes below a certain size, for example, have been exempted from some aspects of regulation; this has almost certainly affected the equipment choice of local service airlines (and of trunk lines on certain routes). The weight limit has been 12,500 pounds gross takeoff weight, with planes below this limit subject to less stringent rules. The Civil Aeronautics Board (CAB), as of January 1, 1969, raised the weight limit to 27,500 pounds, but for jet aircraft only, which discourages further development of more efficient piston, and especially turboprop, aircraft.

Direct regulations setting safety and noise standards also constrain and condition technological change. The effect is not as potent as it might seem, since regulators have had a tendency to set regulations of this type with an eye to what is technically *and* economically feasible. Thus, they may speed the adoption of new technology, but not its development. It was pointed out at the conference that the Federal Aviation Agency (FAA) and NACA (now NASA) have undertaken a good deal of R&D in order to improve technology in safety and other aspects of aviation. Thus the government has been responsible for changes in these areas, though not primarily through regulation.

Phillips suggested that one way to characterize CAB regulation of the airlines is to say that the industry has been allowed to act as a legalized price-fixing oligopoly, with competition diverted to nonprice factors such as speed, range, schedule convenience, and such amenities as attractive hostesses, free cocktails, and movies. Yet regulation has conditioned the structure of the industry through route certification, so that the extent to which the major trunk lines face one or more rivals on important route segments is itself a result of regulatory decisions. Thus, if instead of certifying two or three lines to fly between the same city-pairs the CAB had permitted many more "local" monopoly situations, the history of the rapid adoption of newer aircraft might read differently. In brief, regulation that effectively prohibited price competition and dictated the structure of the air transport industry had an important influence on the pace and pattern of service innovations. Finally, the requirement that any new plane be certified by the FAA as meeting certain minimum performance and safety standards has affected the precise nature of innovation. Certain risks that might have been taken in the drive for speed and low-cost operation have undoubtedly been prevented.

Surface Transportation Industries

Aaron Gellman's chapter suggests a number of specific ways in which regulation by the Interstate Commerce Commission may have significantly affected the pace and pattern of technological change in surface transportation. The ICC's jurisdiction extends over oil pipelines, freight forwarders, and rail, truck, and water transport.

The ICC has had a greater impact on the rate and type of technological change, especially in the railroads, than the FCC and the CAB have had on the industries under their jurisdiction. This may be because the ICC faces a very different regulatory problem than do the other regulatory commissions. In regulating the railroads, the ICC has not had to be concerned with overall levels of profit—the central focus of the Westfield analysis—because since the 1920s the railroads have generally had earnings well below the conventional standard of a fair return on a fair value. (In contrast with the discussion of the communications and electric power industries, the Averch-Johnson effect was barely mentioned in considering the Gellman paper.) For other sectors, the ICC has applied the "operating ratio" method, which ignores rate of return on rate base. Rate regulation as administered by the ICC has been heavily concerned not with profit limitation, but with both intermodal and intramodal competition and with the impact of transport rates on particular shippers and localities.

Minimum rates, established by the ICC because of concern with intermodal competition, have had a disincentive effect on innovation by the rail carriers, according to Gellman, by denying a carrier the right to reduce rates when new, lower cost equipment is introduced. The ability to attract additional business through a rate reduction may be significant in the carrier's decision to incur the risks of innovation—although the practical importance of this effect was questioned. Gellman suggests that the ICC appears to resist proposed rate reductions because of the possible adverse effects on other regulated firms.

In addition to minimum rate regulation, the conference discussion focused considerable attention on the value-of-service principle applied in setting railroad rates. This approach was used before the ICC was established in 1887 and was continued and reinforced by the commission. Under this principle, rates on particular commodities vary according to demand conditions rather than relative costs. High rates are charged for carrying commodities of high value, on the ground that high rates will not discourage their shipment. Rate regulation based on this principle, it was strongly suggested, has a deterrent and distorting effect on innovation because the rate structure is not cost based. A specific illustration mentioned was the "unit train." Part of the resistance to its introduction to move coal from the mines of Appalachia to the eastern seaboard was based on the possibility that it would undermine value-of-service rate making. The very much lower cost at which the unit train could move coal from the mines to the consumer—usually a major electric generating station—could not readily be translated into lower rates without disrupting the rate structure.

As noted above, surface transportation, unlike the communications and electric utility industries, is competitive both between and within modes. Competition has a very special effect on incentives and rewards for innovation, at least when regulation does not interfere. Although the ICC has no direct control over standards of equipment or service (for example, a railroad is relatively free to buy any kind of new rolling stock), nonetheless it can, through the rate-setting process, indirectly exert a significant influence. In setting rates, and particularly in acting on requests for revised rates predicated on the introduction of new equipment, the commission has clearly been sensitive to potential loss by other carriers in the same mode, as well as by carriers operating in different modes. Gellman cites the difficulties encountered in introducing the Big John railroad car to illustrate this point—difficulties that were compounded, he suggests, by the considerable power of the rail carriers themselves to discipline other carriers, as, for example, when an innovation is attempted by one carrier which, rightly or wrongly, is not welcomed by others. In brief, Gellman concludes—and participants in the conference generally agreed—that regulation of surface transportation in this country has had an adverse effect on technological change. This effect has come both directly from rate regulation and indirectly from its impact on the structure of the industry.

In Gellman's paper, and in the conference discussion, the focus was on technological change in the railroads. In addition to the impact of ICC rate regulation, the structure of the industry and the practices followed by the railroads received considerable attention. Although the railroads are to a degree in competition with each other (as well as with competing modes), they are also necessarily interconnected in providing service. Shipments must frequently move over several railroads to reach their destinations; efficient handling of these shipments is possible only if the components of the total system, made up of a number of independent railroads, are integrated. System integration requires institutional and regulatory arrangements to facilitate and promote effective interrelations among the railroads. So far these have been lacking. Some conferees believe that this failure has greatly affected the pace and pattern of innovation. Gellman suggests that the structure of the industry and the attitudes of management (conditioned partly by the regulatory environment of the last eighty years) have led not only to too little innovation, but also to "wrong-headed" innovation.

The handling of freight cars was used as an example of how cooperation among railroads discourages innovation. Freight cars are bought and owned by a single railroad, by a group of railroads, by car-owning companies (as in the case of oil tank cars), or by shippers themselves. Since freight cars move from one railroad to another, a decision by any railroad to acquire a new kind of car that promises to be more efficient or that provides a new or improved service is a decision that affects the whole industry. Often, if one railroad is to make a successful innovation, other railroads have to modify their practices. There are also financial disincentives to innovation. While a freight car is being operated by a railroad other than its owner, a per diem rental charge is paid to the owner. Unfortunately, the basis on which per diem charges are determined does not sufficiently reward the upgrading of equipment. Indeed, until recently, per diem charges did not reflect differences in the age and condition of cars. This fact, together with the whole set of arrangements regarding movement of cars through the system, has provided formidable barriers to innovation. Either some of the benefits are never realized because other railroads fail to modify operating practices or they are realized by other railroads that bear little of the cost. The ICC has taken the position that this set of arrangements among the railroads is not a regulatory matter and has done nothing to encourage improvement in the situation.

A participant suggested, as another example, that the lack of integration is particularly discouraging to innovation. Major yards, especially at key terminals, should in principle be designed to improve traffic flow throughout the country. But since each yard is owned by one railroad (or in some cases jointly by several), yards tend to be designed to promote the efficiency of the owners. Thus managements have had no incentives to use the most efficient techniques for breaking up trains, sorting cars, and assembling trains. Nor have they been motivated to innovate yard equipment that would speed up the rail transportation process. While Gellman recognizes that these problems exist, he downgrades the importance of system optimization over suboptimization by individual railroads. He believes that if only the railroads would do a better job of acting in their own interest most of the inefficiencies in railroad operation would disappear.

The rapid growth of the motor carrier industry is one indication of the serious inefficiencies (including distorted and ineffective innovation and technological change) in surface transportation, in part promoted by the form and type of regulation of the industry, and in part resulting from the management decisions of the railroads. Motor carriers have made heavy inroads into kinds of traffic—bulk commodities over long distances, for example—whose basic cost situation should give the railroads a significant advantage. The ICC's reluctance to allow the railroads to meet the challenge of the trucking industry's improved service by reducing rail rates was cited by some participants as an example of how rate regulation encourages uneconomic technological change. But Gellman emphasizes that the railroads have not taken advantage of even the limited opportunities regulators have permitted.

The conferees agreed that several other interacting factors contributed to poor performance, especially by the railroads. These include the tax treatment of various kinds of expenditures (for example, the limitations on depreciation of the infrastructure). More important in explaining the difficulties of regulating surface transportation without inhibiting socially desirable innovation are the very different ways in which the infrastructure is provided in the various modes. A railroad owns the right-of-way and provides the track. Trucks move on public highways, and although they pay a user charge in the form of fuel taxes, the charge as levied does not provide an incentive to truckers to consider trade-offs between the highway and the truck in minimizing total system costs. (For example, the incentive for the trucker to increase sizes and weights of trucks is natural and is *not* offset by any direct economic effect on him of causing more rapid deterioration of highway surfaces.) Water carriers enjoy virtually a free ride on waterways that are developed and maintained at public expense.

Gellman suggests that concentration in the industry producing rail equipment has further discouraged and distorted innovation. But the significance of concentration was questioned by some, who pointed out the concentration in the electric equipment and the telecommunications equipment industries. Gellman agreed that the attitudes and practices of the railroad unions constitute a very important factor (not discussed in his paper) in explaining the pattern and rate of technological change in the industry. Action by the railroad brotherhoods, such as their insistence on maintaining crew sizes on the railroads despite technical improvements that make reductions possible, has tended to discourage innovation. To the extent that unions have been successful in this kind of resistance, the railroads have not been able to enjoy the reduced costs made possible by new equipment.

A perplexing aspect of all this is that other nations with very different institutions and very different approaches to regulation also have had an unfortunate record in the surface transportation industry—although Canada, with two railroads, is generally given high marks for its transport system. (It was suggested by one participant that this success is partly because each Canadian railroad also operates truck and air lines.) This raised the question, Is there something inherent in transportation activities and technology that makes the development of progressive and efficient industries extremely difficult? No one at the conference was willing to hazard an answer, but all agreed that it belongs on the research agenda for those concerned with the transportation industries in the United States and with the regulatory process and its effects.

Three changes that might improve the situation were considered briefly. First, considerable interest was shown in the possibility of introducing more competition—in other words, of reducing the inhibitions on competition currently resulting from ICC regulation. While proposals for deregulation of transportation to a greater or less extent are not new, some participants emphasized its importance in encouraging innovations and improving the pattern of technological change.

Second, the conferees suggested (as in discussing the FCC) that the range of professional skills and competence of the ICC staff is inadequate to deal effectively with many of the issues facing the commission. Some participants assessed the intellectual environment at the agency as not conducive to the growth and exercise of professional skills. Gellman suggested that the unusually large size of the ICC may explain some of the agency's ineffectiveness. The ICC has eleven members, while the other regulatory agencies have five to seven, and until January 1970, additional difficulties were posed by the annual rotation of the chairmanship.

Third, the question was raised whether these industries might perform better if transportation companies were permitted and encouraged to integrate across modes. Gellman's view, shared by other participants, was that this might lead to a significant improvement in performance.

Conclusion

The conference on technological change in regulated industries was unusual in one important respect. Instead of focusing on a single regulated industry or a particular regulatory agency, the conference examined a problem common to all regulated industries: how to encourage the proper speed and composition of technological change. Virtually all of the literature in industrial organization falls into one of two categories: purely theoretical analyses or empirical studies of a particular industry. Inevitably the industry specialists find numerous shortcomings in the work of the theorists, citing a long list of special characteristics of their own industry that severely limit the applicability of the theoretical models. The chapters in this book can also easily be placed in one of the two pigeonholes.

The reason for this separation within industrial organization is that the technological underpinning and organizational structure of different industries are quite dissimilar, making considerable specialized expertise highly useful, if not absolutely necessary, for understanding the behavior of a particular industry. The depth of specialized knowledge exhibited in each of the industry-oriented papers in this book is impressive and is generally regarded as a prerequisite for evaluating the performance of each industry.

Nevertheless, much can be gained when detailed knowledge of several industries can be combined to paint a broader picture of the behavior of American industry. The conferees, primarily either theorists or experts in a particular regulated sector, spent much of the time at the conference searching for the similarities and key points of difference among the various regulated sectors and between theory and reality. The result was not, nor was it intended to be, a series of specific proposals for improving regulation, nor did a general theory of regulation emerge from the discussion. The conferees did identify several lacunae in the theoretical and empirical research on regulation; these findings are summarized here.

The Performance of Regulated Industries

Nearly all of the participants at the conference believe that the performance of regulated industries falls far short of the ideal and even of a reasonable target for public policy. But they also agree that only in a few exceptional instances can the inadequacy of the performance be clearly documented. Part of the problem in establishing a good estimate of the inefficiencies in the regulated sectors lies in the enormous difficulty of constructing a convincing counterfactual norm. Since no evidence is available on the possible performance of a regulated industry in the absence of regulation or in the presence of regulation in a different form, the norm by which the performance of the industry is measured must be conjectural. The problem is further compounded by the fact that one of the factors that give rise to regulation—natural monopoly arising from capital-intensive economies of scale—also tends to be intimately connected with a high potential for relatively easy and rapid technological advance. The above-average technological progress in regulated industries is a powerful counterargument, at least in the eyes of politicians and regulatory commissions, to theoretical arguments that regulated firms are less progressive than they should be.

Having recognized the difficulties, the conferees went on to agree that the fault lies not only in the stars. Only recently have some important possibilities for comparative research been recognized. One opportunity arises from the fact that most regulation is performed by nonfederal agencies whose quality, regulatory techniques, and statutory responsibilities vary considerably. State and local regulation of telephone systems, power retailing, and transportation offer numerous, largely unexplored opportunities for comparing the performance of similar regulated firms in different institutional environments.

Theoretical work on the expected behavior of the regulated firm could be more useful if based on a more relevant comparative norm than the conventionally used unregulated monopolist or perfect competitor. Several conferees thought the Westfield paper an important, pioneering effort because the author tried to analyze the sensitivity of the performance of regulated industries to different regulatory rules.

A minority of the conferees offered the opinion that even comparative theoretical and empirical research would not be particularly useful because, they contended, the fact of regulation is a much more pervasive influence on firm behavior than the technique of regulation. Most of the conferees agreed that the existence of regulation can always be expected to create important social costs, but that the magnitude of the costs is significantly affected by the choice of regulatory techniques. If regulation, per se, does create certain important inefficiencies, research into the proper scope of regulation is of utmost interest.

The Scope of Regulation

The conference discussion often centered around the issue of defining the boundaries of regulation. Contrived scarcity by a natural monopolist has not been the only circumstance to give rise to regulation, much of which has been anticompetitive from the beginning. Many services normally regarded as natural monopolies have remained natural monopolies only because of the interaction between regulation and technological change.

Only occasionally has research been relevant to the problem of defining the proper scope of regulation. Estimates of the costs of regulation, for example, while only crudely approximated thus far, are essential to deciding whether attempting to cure a specific market inefficiency through regulation is worthwhile. No attempt has been made to assess the extent to which the cost and the effectiveness of regulation differ according to the scope of the responsibilities of the regulatory agency. Should single agencies be given the responsibility of regulating entire sectors, or should each industry in a sector be regulated by a separate agency? Are regulators more effective at regulating profits, prices, or qualitative performances? How is the proper scope of regulation affected

by the organizational and technological characteristics of the industry being regulated?

The Regulatory Process

Perhaps the most neglected topic in research on regulation is the behavior of the regulatory agency. Economists are given to assuming that
the purpose of regulation is to limit the profits and prices of natural monopoly with as little resulting inefficiency as possible; the extent to which
a regulatory agency falls short of achieving this objective is regarded as a
measure of its incompetence. While this view may be a valid description
of how regulation should, in principle, operate, the conference discussion
showed that it has little basis in the legislation creating regulatory agencies or in the actual behavior of the agencies. In addition to the considerations of efficiency that dominate the economics literature, the conferees
agreed that regulators are also concerned with the health of regulated
firms, with the income-distribution effects of industry price structure and
composition of service, and with the politically, as well as economically,
efficient degree of service reliability. Many participants in the conference
believe that regulatory agencies tend to avoid risks, to "crawl in bed with
the industry" they regulate, and to take actions that favor well-defined
and well-organized political pressure groups (such as the rural lobby),
but these are only impressions, the existence and importance of which
have yet to be verified conclusively. Regulatory agencies obviously have
failed to consider many of the social costs and benefits of the regulated
industry that are external to the agency and the firms, such as environmental considerations.

If the regulatory agency is to be regarded as motivated by a complex
amalgam of often mutually inconsistent objectives, a number of uninves-
tigated and highly important research topics remain. Do the objectives of
regulation, and the effectiveness of agencies in achieving these objectives,
vary systematically with differences in the organizational structure of the
agency? How does the behavior of commissions differ according to the
size and composition of the commission? How do the goals and performance of agencies with legally oriented staffs and adversary hearing procedures differ from those of agencies more oriented toward economics
and neutral fact-finding? How does the actual performance of the industry affect the goals and policies of the agency, and does the nature of this
interaction also vary systematically with the structure of the regulatory

system? What constraints on the behavior of regulatory agencies might partially correct for some of the less desirable objectives, actions, and inefficiencies that may be inherent in the regulatory process?

In the discussion of each of the regulated sectors, a persistent theme was the reluctance of the regulatory agency to alter the structure of the industry it regulates. Most often this reluctance is manifested in anticompetitive actions, preserving protected monopoly positions of existing firms, but during the conference several instances were cited in which the regulatory agencies had balked at potentially beneficial concentration. Within the communications sector, the FCC has not given serious attention to the efficiency gains that might be attained by opening up equipment manufacture to competition. At the same time the commission has shown little interest in the "single entity" proposal for international communications and has prevented the entry into broadcasting of regional radio and television stations in order to protect the franchise of local stations, even though in the case of network television stations programming is virtually identical on stations in adjacent geographical areas. In transportation, the ICC has looked more favorably upon the clearly anticompetitive mergers of parallel rail lines than upon intermodal or connecting line mergers, despite efficiency arguments which seem to support precisely the opposite policy.

Research into the regulatory process and the behavior of regulatory agencies might yield fruitful insights about how to make regulation more flexible and less enamored of preserving the status quo. One suggestion made at the conference was that regulatory agencies be given the authority to compensate the victims of changes in regulatory policy. While compensation has led to serious problems in the instances when, for different reasons, it has been tried (such as subsidies in the maritime and air carrier industries), most conferees agreed that the idea was worth serious consideration.

Regulatory Techniques

The impact of various regulatory rules and procedures cannot be assessed without considering the intentions of regulators; consequently, definitive conclusions about the actual and potential power of regulatory devices cannot be reached by observing the actual performance of regulated firms until the behavior of the regulatory agency is better understood. Nevertheless, tentative hypotheses were expressed by conferees on the importance of several types of regulation.

A majority of the conferees believed that of least importance in the long-run development of regulated industries have been the controls placed on the general price level, such as limiting the rate of return. While rate-of-return regulation was regarded as probably creating a bias toward capital-intensive technology, most conferees believed this to be far more important in affecting static efficiency decisions than in influencing technological change. Several conferees suggested that detailed price regulation, such as that practiced by the ICC for railroads, has been extremely damaging to progressivity, although they agreed that, with different motivations, a regulatory agency might be able to use detailed price regulation very effectively to achieve abnormally progressive behavior.

Commanding wide support was the view that the pace and pattern of technological change is primarily determined by a few, discrete regulatory decisions on matters related only tangentially to prices and profits. For example, decisions to permit a new firm to enter an industry or to allow a new technology to be exploited have profound effects on a regulated sector.

The conferees also recognized that regulatory agencies employ numerous mechanisms for controlling regulated firms which are rarely analyzed in depth but which are possibly extremely important. One such policy instrument discussed at length in this book and at the conference is the ability to control mergers and intersystem cooperation. Another is the cost imposed on regulated firms by the regulatory process—a cost that can be avoided through informal acquiescence to the desires of the staff or commissioners of an agency. Still another is control over which investments can be included in the rate base: the FCC and the FPC have used this instrument to influence the investment plans of regulated firms. Still another is the control exercised by the regulatory commission over product quality.

Future Directions in Regulatory Policy

The participants in the conference reaffirmed the strongly held view of many academics and public officials that the regulated sector of the American economy is not performing as well as it could with more enlightened public policy. The conferees generally agreed that the whole approach of public policy in regulated sectors needs to be reexamined, from redefining the purposes of regulation to reassessing the effectiveness of the regulatory techniques commonly employed.

The conferees did not reach a consensus on many of the specific questions of regulatory policy and practice raised at the conference. The tenor of their discussion was that on most of these issues much difficult, relevant research must be done before widespread professional support can be won for specific changes. (A notable exception was the near unanimous view that the ICC's method of calculating rail freight charges, based on historical average costs and "value of service," ought to be significantly altered, if not abandoned, to bring prices more in line with actual costs and thereby to remove a substantial impediment to efficiency.) The conference did succeed in pointing out the likely directions in which policy changes should be made and the research topics that need to be investigated before more than a qualitative sense of the direction of potentially fruitful changes in regulatory policy can be gained.

Conference Participants

WILLIAM M. CAPRON *Brookings Institution*
RICHARD E. CAVES *Harvard University*
S. DAVID FREEMAN *Office of Science and Technology*
AARON J. GELLMAN *Budd Company and University of Pennsylvania*
ABRAHAM GERBER *National Economics Research Associates, Inc.*
KERMIT GORDON *Brookings Institution*
ORRIS C. HERFINDAHL *Resources for the Future*
WILLIAM R. HUGHES *Boston College*
LELAND L. JOHNSON *RAND Corporation*
NICHOLAS JOHNSON *Federal Communications Commission*
JOHN V. KRUTILLA *Resources for the Future*
HARVEY J. LEVIN *Hofstra University*
PAUL MACAVOY *Massachusetts Institute of Technology*
JOHN G. MCGOWAN *Yale University*
JAMES W. MCKIE *Vanderbilt University*
H. MICHAEL MANN *Department of Justice*
EDWARD MARGOLIN *Interstate Commerce Commission*
EDWIN S. MILLS *Johns Hopkins University*
JAMES R. NELSON *Amherst College*
RICHARD R. NELSON *Yale University*
ROGER G. NOLL *California Institute of Technology*
JOSEPH A. PECHMAN *Brookings Institution*
MERTON J. PECK *Brookings Institution*
ROBERT H. PERRY *RAND Corporation*
ALMARIN PHILLIPS *University of Pennsylvania*
FREDERIC M. SCHERER *University of Michigan*
SAM H. SCHURR *Resources for the Future*
WILLIAM G. SHEPHERD *University of Michigan*

227

RICHARD SOLOMON *Federal Power Commission*

IRWIN M. STELZER *National Economics Research Associates, Inc.*

JOHN E. TILTON, JR. *Brookings Institution*

HARRY M. TREBING *Michigan State University*

HASKELL P. WALD *Federal Power Commission*

FRED M. WESTFIELD *Vanderbilt University*

LEE WHITE *Federal Power Commission*

OLIVER E. WILLIAMSON *University of Pennsylvania*

GEORGE W. WILSON *Indiana University*

Note: The affiliations given for the participants listed above are those of February 1969, when the conference was held. As of July 1970, the following participants had changed their affiliations: William M. Capron, Harvard University; William R. Hughes, Charles River Associates; Edwin S. Mills, Princeton University; Roger G. Noll, Brookings Institution; Merton J. Peck, Yale University; John E. Tilton, Jr., University of Maryland; Lee White, Semer, White, and Jacobsen.

Index

Index

surface transport, 168, 174–75, 179–82, 186, 194–96, 215–20

Telecommunications system. *See* Communications industry

Telegraph, 89, 98

Telephonic communications: economies of scale, 3–4; effect of regulation on, 11, 98–100; FCC investigation of rates, 116; history of, 89–90; monopoly in, 42, 94–95, 118*n;* need for research in, 221; system integrity in, 6–7, 23. *See also* Communications industry

Teletype Corporation, 110*n*, 118*n*

Teletypewriter service, 116

Television, 110, 113–14, 224

TELPAK, 114, 116–17

"Tenancy in common" in electric power industry, 55*n*, 58

Tennessee Valley Authority (TVA), 55, 76*n*

Terminal devices in telecommunications, 89, 91, 95–96, 103, 109–11, 208

Thermal efficiency, 61–64, 68, 82

Tobin, James, 88*n*

Trans World Airlines, 140, 146

Transcontinental and Western Air, 140–43, 147, 150*n*, 151*n*, 152, 155–56

Transmission: in communications system, 87, 89–90, 95*n*, 100, 114–15; of electric power, 45–50, 59, 61–62, 65*n*, 71–73, 201–03

Transportation. *See* Air Transport industry; Surface transport

Transportation, Department of, 12

Trebing, Harry M., 59*n*, 117*n*

Troxel, C. Emery, 102*n*, 168*n*

Trucking, interstate. *See* Highway transport

Unions, labor, 29, 219

United Air Lines, 140, 142, 144, 145*n*, 146–47, 151*n*, 154–57

United Kingdom, 68*n*, 101*n*, 104*n*, 112*n*, 153

United States: electric power industry, 46, 53, 68*n;* telecommunications industry, 90–91

Usher, Dan, 87*n*

Utilities, public, 45*n*, 53, 75*n*, 76*n*, 78–80, 82

Value-of-service principle in railroad rates, 216, 226

Vertical integration: in air transport, 211; in communications industry, 98, 101, 104*n*, 111–12, 122; in surface transport, 187–90

Viner, Jacob, 23

Wages, 6

Water transport, inland, 2, 167, 181–82, 191, 215, 219

Weber-Fechner law, 32

Western Air Express, 140, 144, 155

Western Air Lines, 142, 146–47

Western Electric Company: antitrust action against (*1949*); 100; efficiency of, 104*n;* equipment production by, 90–92, 95*n*, 98, 102, 107, 110; FCC's consent decree about, 206–07; invention in, 86; need for further study about, 118, 120; rate of return in, 112

Western Union, 98, 100, 104–05

Westfield, Fred M., 12, 19*n*, 20*n*, 21*n*, 80*n*, 106*n*, 197–201, 215, 222

Whitnah, D. R., 144*n*

Wholesale-retail specialization in electric power, 83, 85

Williamson, Oliver E., 104*n*

Wolbert, George S., Jr., 182*n*

Yarbrough, John F., 178*n*